WEDDELL SEAL, CONSUMMATE DIVER

WEDDELL SEAL

consummate diver

GERALD L. KOOYMAN

Physiological Research Laboratory, Scripps Institution of Oceanography

CAMBRIDGE UNIVERSITY PRESS

Cambridge

London   New York   New Rochelle

Melbourne   Sydney

CAMBRIDGE UNIVERSITY PRESS
Cambridge, New York, Melbourne, Madrid, Cape Town, Singapore, São Paulo, Delhi

Cambridge University Press
The Edinburgh Building, Cambridge CB2 8RU, UK

Published in the United States of America by Cambridge University Press, New York

www.cambridge.org
Information on this title: www.cambridge.org/9780521112413

First published 1981
This digitally printed version 2009

A catalogue record for this publication is available from the British Library

Library of Congress Cataloguing in Publication data
Kooyman, Gerald L
Weddell seal, consummate diver.
Includes index.
1.   Weddell seal.   2.   Mammals – Antarctic regions –
McMurdo Sound.   I.   Title.
QL737.P64K66   599.74′8   80–18794

ISBN 978-0-521-23657-7 hardback
ISBN 978-0-521-11241-3 paperback

Contents

Preface

In recent years, several fine books have been published that are devoted to comprehensive, life-history accounts of single, terrestrial, mammalian species. The major tools for these studies were notebook, binoculars, and keen observation.

The complexity of the sea and the size of some marine mammals present insurmountable obstacles to comprehensive studies. Underwater visibility in the sea is seldom good enough to see from one end of some of the larger whales to the other; obviously, the observational tools in the sea, as in most natural situations, must be complex. However, there is one place where an unusual set of circumstances has reduced the complexities of such studies and provided us with a window into the underwater world. It is one of the most isolated wildernesses on earth: McMurdo Sound, Antarctica. During most of the year, this is the exclusive kingdom of the Weddell seal. Nowhere else, not even in other parts of Antarctica, is it possible to observe and study Weddell seals or any other sea mammal in such detail as here. Yet, paradoxically, the characteristics that make this place so suitable for scientists to observe seals also make it one of the most difficult places for a sea mammal to survive. What we have learned about the Weddell seals in this area seems to justify this contention. The information obtained at times borders on the sensational. By extrapolation, it also has given us new insight into other marine mammals and how extraordinary their life patterns must be.

It is my goal in this book to give the lay public, undergraduate and graduate students, and my peers a comprehensive description of some of the remarkable characteristics and habits of this species; and to do it within a single work. This has not been done before. In my effort to satisfy such a broad audience, I hope that the book will not seem overly detailed in some cases, or so simple in others that readers will lose interest in the whole. If my objective is achieved, then I feel that not only will the public's world be enlarged, but some part of my debt to Weddell seals will be repaid.

I am indebted to many for assisting in the work, and for giving advice and support. Financial aid that made the various studies possible has come from the Division of Polar Programs of the National Science Foundation and from the Heart and Lung Institute of the National Institutes of Health. The former is also responsible for a tremendous logistic program, which makes all field projects possible. The people who have helped me over the more than fifteen years of my fascination with Weddell seals are too numerous to acknowledge individually. Some are mentioned in the book. To the others, I say that omission does not reflect any less gratitude on my part. I must make two exceptions: Dr. George A. Llano, who at the inception of my research said it was possible, and Professor Albert R. Mead, who was willing to extend the Arizona horizon to Antarctica.

G. K.

To vivacious, enthusiastic Melba:
Always present in spirit,
but for whom Antarctica has been
a vicarious experience

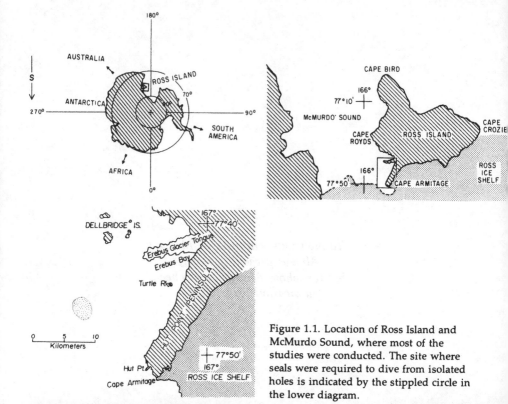

Figure 1.1. Location of Ross Island and McMurdo Sound, where most of the studies were conducted. The site where seals were required to dive from isolated holes is indicated by the stippled circle in the lower diagram.

Figure 1.2. Aerial photograph of the eastern shore of McMurdo Sound taken from a helicopter at 3,000 m and facing north. The prominent peak is Mount Erebus, an active volcano approximately 4,000 m high. (Photograph by Dan Costa, taken January 1979.)

1

McMurdo Sound

No latitude for error.
Sir Edmund Hillary

The stage for much of the drama I shall describe is McMurdo Sound, Antarctica. To the west is the continent; the Queen Victoria range towers above the shoreline. To the east, the Sound is bounded by Ross Island, derived from three volcanic cones and dominated by still-active Mount Erebus at 4,000 m. Locked between Ross Island and the continent, this small sea is protected from the major currents of the Antarctic Ocean. As a result, an ice sheet covers the southern portion of the Sound for most of the year (Fig. 1.1 and Fig. 1.2).

Large areas of the Sound begin to freeze as the autumn temperatures decline, usually in late April. By August the mean monthly temperature is −26°C, with daily minimums commonly as low as −40° to −50°C. At this time, the ocean surface is covered for miles with a layer of sea ice several feet thick that is interrupted in only a few places. A break in the ice always occurs along the shoreline, where the daily tidal changes continually raise and lower the level of the sea and prevent the bonding of land and water. Away from the shore, there are only scattered cracks, many of which freeze shut soon after they appear. Others, due to peculiarities in winds and ocean currents, remain open (with only a thin layer of ice) all winter. Along these cracks where the ice is thin, the Weddell seals (*Leptonychotes weddelli*) keep their breathing holes open during the winter. A seal breaks thin new ice with a blow from the top of its head; if too thick for that method, the seal cuts the ice open with its upper canines and incisors.

From a helicopter, these half to one meter diameter black breathing holes are easily seen scattered over the icy plain. To a Weddell seal, its hole is critical. It is the seal's link to the sea and food, and after a dive its link to the surface and life-sustaining oxygen. How does a seal perceive its hole when on top or below the ice? When on top of the ice, its angle of view is very oblique, and even from a short distance the hole cannot be seen. Such features as ice hummocks around the cracks are probably important aids. Also, the seal seldom wanders more than three meters from its hole.

The seal's problems under the ice are much different. Its ability to see distant objects is enormously reduced, even though these waters are probably the clearest in the world. In the summer, the holes appear as brilliant beacons when the sunlight beams through. In the winter, when the sun does not rise for three and a half months, the beacons are out. Furthermore, a seal can hold its breath for only a finite time. Therefore, during both summer and winter the seal must be able to navigate with pinpoint accuracy, and it must not run out of oxygen before it reaches its destination. The seal's skill in and methods of under-ice orientation together form one of the most fascinating problems in animal behavior. But rather than get ahead of my story, the setting of the stage should be completed.

Although the long winter nights and low temperatures that bring about the freezing of the sea seem severe at McMurdo Sound, they are relatively mild compared to the conditions of other parts of Antarctica. At the South Pole, the sun sets for six months of the year, from March 20 to September 20. Temperatures in the coldest known part of Antarctica are at the 3,488 m high Vostok Station, which is located well inland on the polar plateau. Here, temperatures sink to a monthly mean in August of −68°C. The lowest observation was −88°C, which was recorded August 24, 1960. This is the lowest, naturally occurring ground-level temperature on this planet. In comparison, the coldest month in the Arctic averages only −35°C, with a record low of −68°C.

The temperatures in the Arctic and at McMurdo Sound are not so extreme because of the lower altitude and because of the ameliorating effect of the sea. At McMurdo, the sea temperature is unusually uniform, averaging −2.0°C year round. The annual temperature range is less than 1.0°C, and there is almost no variation with depth. The temperature changes less than .5°C, ranging from the surface to 275 m. In contrast, the temperature range of the sea off San Diego in the summer may be as much as 11.0°C between the surface and 30 m.

Temperature is not the only condition that is less severe in antarctic waters. Ocean currents are rarely more than 3 or 5 km per hour, but the average surface wind at Hut Point, which is in the most southern sector of McMurdo Sound, is 24 km per hour. Possibly the windiest place in the world is Cape Denison, which Douglas Mawson's overwintering party sadly learned. The average annual wind speed was 73 km per hour.

There is very little continental shelf surrounding Antarctica; the water becomes deep near shore. Only five or six km from Hut Point the depth is 600 m. The nearest large land mass is South America, the tip of which is about 900 km from the tip of the Antarctic Peninsula. Looking at these two fingers of land on a globe, they appear to be reaching to touch one another as they once did millions of years ago. Now they are separated by one of the wildest seas in the world. Everywhere else around Antarctica this great southern ocean separates the continent from other significant lands by more than 1,600 km. Such a formidable barrier and such unwelcome weather prevented an invasion by land animals. There are no polar bears or Eskimos; indeed, the most significant residents on this fifth largest continent (33 million square km) are a few insects and mites, the largest being less than 1.6 mm long. In contrast, the sea is rich with life and is as productive as any sea in the world.

If it were not for scuba equipment and deep sea underwater cameras, we would be rather ignorant of the context in which the animals of the sea live. The usual method of collecting data about the bottom, for example, is either to lower a grab or dredge the floor and examine the specimens brought up in this equipment, or to set baited traps for the same purpose. Can you imagine describing a tropical rain forest by similar random-sampling methods? What kind of an impression would we have of a forest if we had never seen it, and all the information about it had been gathered by piecemeal sampling from several hundred to thousands of meters above it without ever being in the forest itself? Fortunately, technological achievements have increased our mobility and observational skills, and our understanding of the sea is broadening.

For the last few years, scientists using scuba equipment have begun to describe in detail the shallow bottom communities of McMurdo Sound. An intensive study of this kind has been carried out by P. K. Dayton and G. Robilliard near Hut Point (Dayton, Robilliard, and Paine, 1970). At the time, these two were graduate students at the University of Washington. During an intensive period of study between October 15 and December 11, 1967, they made nearly 200 dives. They followed a similar schedule in 1968. Some of these dives lasted one and a half hours, and in some cases they descended to 60 m. They have described three zones down through these depths.

Zone I extends from the surface to 15 m (Fig. 1.3a). It is a rather sterile area, the least inhabited, because ice forms on the bottom. It does so in the form of large, beautiful crystals, which accumulate into mats up to 0.5 m thick (Fig. 1.3b). As these crystal clusters grow, they become more buoyant because the specific gravity of ice is lower than that of water. In time, depending upon how well-anchored the ice is, it will break free and rise until it comes against the underside of the sea ice. When "anchor ice" lifts from the bottom, it frequently carries with it animals that have become frozen in it. In some instances, objects as heavy as 25 kg have been carried away. The result is that almost all sessile, or stationary, animals that attempt to settle in

Figure 1.3. (a) This remarkable seascape photograph of zone I was a time exposure taken at a depth of 20 m and facing toward the surface. The slit of light in the right-center is a break in the ice 245 m away. The picture was taken in October, a period of maximum water clarity. (Photo by John Oliver.)

this zone are doomed. (Rather interesting is the observation that many of the sea urchins and starfish that become frozen in the anchor ice recover from this state and are able to resume normal activities when the ice melts.)

Zone II ranges from 15 to 30 m below the surface. It consists of a cobble and lava substrate upon which many sessile and motile animals are found (Fig. 1.4). Anchor ice also forms in this zone, but in a much more dispersed manner so that few sessile animals are affected.

Zone III begins 31 m below the surface; its lower limit is unknown. The bottom is covered with a thick mat of siliceous sponges, which is sometimes over 1 m thick (Fig. 1.4). It is rich in animals living on its surface and within the mat. It has been estimated that there are more than 12,500 animals per liter of mat substance. The upper limit of this zone is the lower limit of anchor ice formation. Why the ice stops forming at about 30 m is unknown.

Figure 1.3. (*b*) In this shallow water photograph, the anchor ice growing in large clusters on the bottom can be seen in the foreground. The diver just below the tidal crack is carrying an underwater camera.

Little is known about the pelagic, or open ocean, animals that live in the midwater depths. Aside from the zooplankton (the small, drifting invertebrates), the most abundant animal of all is the squid. Few specimens have been collected, yet it is the most important food item of the emperor penguin and, in some areas, of the Weddell seal. Based on the size of squid beaks found in the stomachs of emperor penguins, some of these squid may have been one-third to two-thirds in meter length.

Of the two or three most common species of fish, the antarctic herring, *Pleuragramma antarcticum*, which is similar in size and appearance to the herring of temperate waters, is the most abundant; it occurs from near the surface to several hundred meters below. The next most abundant fish, similar in size, is the ice-loving *Trematomus borchgrevinki*. (In many cases, only scientific names have been given to animals of the Antarctic, and this fish is an example. It was named for Carsten Borchgrevink, the leader of the first land-based, overwin-

Figure 1.4. Seascape of zones II and III. This was a time exposure taken at 30 m and facing down over a bluff. The bottom of the valley is 60 m deep. The white sponge at the base of the valley (arrow) is 1.3 m tall. (Photo by John Oliver.)

tering expedition to Antarctica.) This fish lives under the sea ice near the surface, and it utilizes the crystal layer formed underneath the sea ice as a place to rest and hide.

The largest fish that occurs in antarctic waters, *Dissostichus mawsoni*, was until recently known from only a few specimens that were stolen from Weddell seals when they brought them to the surface to eat (Fig. 1.5). The largest specimens weigh up to 65 kg and are over one and a half meters long. They look almost like grouper or black sea bass. In the last two years, A. L. DeVries of the University of Illinois has devised methods of catching these fish on hooks set in a vertical array from the sea floor to just under the ice (Raymond, 1975). Most of the fish are caught at midwater depths (150 to 300 m), and they appear to be rather common.

The most unusual fish in the McMurdo Sound area is the ice fish, *Chaenocephalus aceratus*, so named not because it lives in the ice, but because it is so pale, and for good reason (Fig. 1.6). It has no red blood cells, no hemoglobin (the blood respiratory pigment), and no myoglobin (the muscle respiratory pigment) (Ruud, 1965). Its gills are white, its heart is pale yellow, and its blood is gin clear. Again, there has been little success in collecting ice fish at McMurdo Sound, except for those taken from Weddell seals.

The birds and mammals that occur consistently in McMurdo Sound can all be counted on the fingers of one's hands. Such a simple ecosystem of birds and mammals is uncommon.

Of the birds, the south polar skua, *Catharacta maccormicki*, sometimes called the eagle of the Antarctic, is the most commonly seen. It is large, strong flying, and rapacious (Fig. 1.7). It frequents the refuse dumps at most of the antarctic bases. This bird has even been seen at the South Pole. Its stay in the area is seasonal, from October to March.

Only two species of penguin are seen in McMurdo Sound: The first is the Adelie penguin, *Pygoscelis adeliae*, which is an average-size penguin standing 76 cm tall and weighing up to 6.5 kg (Fig. 1.8). Its principal foods are fish and krill. (The latter is a shrimplike invertebrate called "krill" by the Norwegians. The two main species of krill in Antarctica are *Euphausia superba*, the primary food of the large whales, and the more coastal *E. crystallorophias*.) The Adelie penguins do not remain in the area during the harsh winter. They spend March to October in the pack ice, that large region of broken sea ice surrounding the antarctic continent. The second is the emperor pen-

Figure 1.5. Antarctic cod, *Dissostichus mawsoni*, the largest fish known to occur in McMurdo Sound. The fish pictured was captured by a seal and taken from it. The carabiner holding the fish is about 8 cm long. This fish was medium size and probably weighed 25 kg.

Figure 1.6. Antarctic ice fish, *Chaenocephalus aceratus*. This fish is unusual in its lack of red blood cells, or any oxygen binding hemoglobin in the blood or myoglobin in the muscle.

Figure 1.7. South polar skua, *Catharacta maccormicki*. The wingspan is 1.3 m and it weighs 1.8 kg.

Figure 1.8. An Adelie penguin, *Pygoscelis adeliae*, being chased ashore by a leopard seal, *Hydrurga leptonyx*.

guin, *Aptenodytes forsteri,* which is the largest diving bird in the world (see Figure 1.9). It stands up to 1.2 m tall and weighs as much as 45 kg. If any bird or mammal is more bound to the southern high latitudes than the Weddell seal, it is the emperor penguin. These birds usually nest on sea ice adjacent to the continent or on nearby islands. They begin to arrive at the breeding sites in April and do not depart until February of the next year. This leaves them only two months to wander at sea before returning for another breeding season.

Four species of mammals, other than the Weddell seal, are commonly seen in McMurdo Sound. Like the skua and Adelie penguin, they are only summer visitors. Almost every summer, killer whales, *Orcinus orca,* and minke whales, *Balaenoptera acutorostrata,* come south into the Sound, using the large open leads and the ice breaker channel made by the U.S. Coast Guard. The minke whales, which

Figure 1.9. *(Left)* An emperor penguin, *Aptenodytes forsteri,* adult in attendance with a group of chicks. *(Right)* Emperor penguin adult diving to depth.

are up to 9 m in length, come to feed on the plankton that is abundant at this time of year (Fig. 1.10). The killer whales, the males of which may be slightly over 9 m in length, are hunting for seals (but not necessarily Weddell seals), because at this time of year a number of crabeater seals, *Lobodon carcinophagus*, are also in the area (Fig. 1.11). Despite their name, crabeater seals feed primarily on krill; there are no crabs in Antarctica. This animal is about 2.7 m from nose to tail tip and weighs between 225 and 270 kg. Many individuals bear the deep scars in their blond coats of previous encounters with leopard seals. The crabeaters' usual range is the pack ice. It is the most abundant of the antarctic seals.

The leopard seal, *Hydrurga leptonyx*, is the most catholic feeder perhaps of all seals (Fig. 1.8). This 3 to 3.5 m long, 450 kg animal feeds on krill, fish, penguins, and other seals. It is mainly an animal of the pack ice, but some individuals come south in the spring and

Figure 1.10. A minke whale, *Balaenoptera acutorostrata*, breaking through thin ice to breathe.

in the summer as well to catch penguins near their rookeries.

Last, but hardly least, of the animals that share the McMurdo Sound area with the Weddell seal is man. The continent and adjacent islands existed virtually uninhabited by humans until this century. Although ancient scientists speculated about a southern continent, and named it *Terra Incognita*, it was not until the eighteenth century that Cook sailed into the southern seas and charted some of the sub-antarctic islands. After that expedition, exploration of the southern continent gradually increased. Now there are ten to fifteen year-round bases in Antarctica. The largest of these by far is McMurdo Station, an American base, on Ross Island. About 200 men overwinter at this station, and in the summertime its population grows to more than 1,000. Also on Ross Island is New Zealand's Scott Base, which has a complement of 20 to 30 men. The primary mission of these bases is to support scientific research programs.

Suggested reading

Mountfield, D. 1974. *A History of Polar Exploration*. New York: Dial Press.
Stonehouse, B. 1972. *Animals of the Antarctic: the ecology of the Far South*. New York: Holt, Rhinehart and Winston.

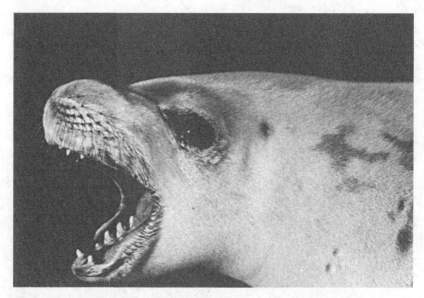

Figure 1.11. Crabeater seal. *Lobodon carcinophagus*, threatening. Note the complex molars, which are presumably used for straining the krill from any seawater taken in during capture of the prey.

2

The Weddell seal

This creature resembles the quadruped of
the same name in being spotted. He
[Jamieson] considers it to be a new
species of phoca.
James Weddell, South Orkneys, January 15,
1823

The Weddell seal was first de-
scribed by Robert Jamieson in James Weddell's book *A Voyage To-*
wards the South Pole Performed in the Years 1822–1824. The description was based on six skins collected at Saddle Island, South
Orkneys, on January 13, 1823. Also in the book is a rather fanciful
drawing of the seal by Captain Weddell, which is captioned "Sea
Leopard of South Orkney's" (Fig. 2.1). Later, based on the illustrations and report in Weddell's book, the animal was officially described for science in 1826 by R.P. Lesson. The Latin binomial he
gave it was *Otaria weddelli*. In 1880, the animal was renamed *Lep-*
tonychotes weddelli by J. A. Allen; this is its official name at present.

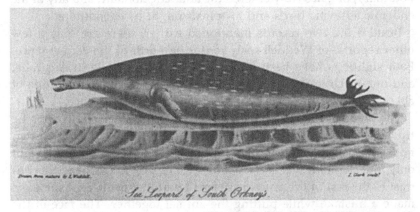

Figure 2.1. One of the first illustrations of
the Weddell seal. (From Weddell, 1970.)

Species descriptions usually require skull specimens as well, but these were not obtained until some years after Weddell's expedition. Oddly, they came from two animals found near the Santa Cruz River in Patagonia, which is near the latitude of 50°S. The species probably has not been seen in regions so far north since. It occurs in greatest abundance near the coast of Antarctica. For example, within a radius of 25 km of Hut Point, a few hundred pups are born each spring. Small breeding populations occur in some of the subantarctic islands, the most northerly of which is South Georgia Island, where, in 1956, twenty-six pups were born.

Interestingly, only one animal has been sighted in the Falkland Islands, which have a latitude nearly the same as South Georgia but, because of the oceanic currents, have a very different climate. South Georgia Island is subantarctic. The mountains are covered with glaciers and pack ice reaches its shores in winter. In contrast, the Falkland Islands are cold, temperate, and have a climate more like that of London, England. The great difference between the two island groups is due to their position relative to the antarctic convergence, which is an oceanic boundary between the Antarctic Ocean and the Pacific, Atlantic, and Indian oceans to the north. The latitude of the convergence varies with the season, but this average falls between 50° and 60°S. At the convergence, the cold, dense waters of the Antarctic sink below waters of the northern regions. As one moves south over this boundary, the sea temperature drops 2° or 3°C, and sometimes there is a change in the color of the water. Once past this boundary, one has truly entered the antarctic domain, and any of the endemic antarctic birds and mammals might be encountered.

Besides the two records mentioned earlier, there are only a few other reports of Weddell seals venturing north of the convergence. Four sightings have been made in New Zealand, one in Australia, and one in Uruguay. The sighting in Uruguay was at a latitude of about 35° S, which is the northernmost known occurrence.

The Weddell seal is one of the largest of all seals. Its length from the tip of its nose to the tip of its flippers is about 3 m. In early spring, both males and females commonly weigh 400 to 450 kg. By weight only, the elephant seal and leopard seal are larger. The head, in contrast to the rest of the body, is so small that it appears as if someone made a mistake while putting the animal together. The face of the Weddell seal is one of the most benevolent in the animal world. Its muzzle is short, giving it the appearance of a juvenile, and its mouth always seems to have the hint of a smile. Its eyes are large and deep brown. The vibrissae (moustache or whiskers) are usually short and undistinguished, as is typical of all the antarctic seals.

The fur, which is about a centimeter long, covers the entire body, except a small portion of the underside of the fore and hind flippers. Shortly after moulting, the back is blue-black, grading to a silver-white spotting on the belly (hence the early name of sea leopard). As the fur ages, it fades. Shortly before moulting, the back is a rust brown.

The skull is distinctive. The bones are thin and light for such a large mammal. However, the canines are robust and project forward, as do the incisors. If it were not for these peculiar dental features, especially in the upper jaw, and the small, narrow snout, the Weddell seal probably would not be able to remain year round in McMurdo Sound (Fig. 2.2).

These dental features are used as a very effective ice ream, which the animal must rely on frequently. During most of the year, air temperatures are so low that the breathing holes will freeze over if the

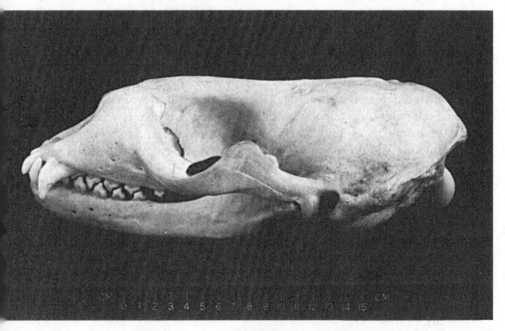

Figure 2.2. A lateral profile of the Weddell seal skull. Note the forward-projecting, robust canines and incisors, the teeth so important as ice reamers.

seal remains away long. When a seal returns to such a hole, its forward-projecting incisors cut at the flat surface. Once a small hole is made, most cutting is done with its upper canines. The seal literally hangs on its canines as it vigorously sweeps its head back and forth, abrading the ice in the process. In such a manner, the hole may be enlarged just enough to permit the seal's head to protrude so it can breathe. Frequently, the hole is enlarged sufficiently to permit its entire body to pass through, enabling it to leave the water.

Studies of the Weddell seal can be divided into three historical periods: (1) the heroic era, (2) the transition period, and (3) the modern era. Only one rather comprehensive study occurred during the heroic era. That was a description of general habits by E. A. Wilson (1907) during the two and a half years that he spent at Hut Point, McMurdo Sound, on Captain R. F. Scott's first expedition to Antarctica. The expedition was away from England from July 1901 to September 1904. This was a time when all antarctic expeditions were long and there were great hardships. The explorers voyaged to the south in ships powered by both coal and sail. The journeys were long, and upon arrival the stations were built by the expeditionary personnel. An important source of food was the fresh meat obtained from butchering local seals and penguins. Overland travel was by dog sled or man-hauling sled, and there was no communication with the outside world.

The transition period was a time when expeditions were still at least eighteen months long. Transport was still by both sail and coal. Seals were occasionally used as food and dog teams were still the main means of land transport, but powered vehicles were coming into use and there was radio communication with the outside world. Three major works were published in this period. The first was A. A. Lindsey's study conducted at the Bay of Whales during the austral summer of 1934–5 (Lindsey, 1937). This was done on Admiral R. E. Byrd's second antarctic expedition. Close on the heels of this study was that of G. C. L. Bertram (1940). This work was carried out during a British expedition to the Antarctic Peninsula between January 1935 and March 1937. Somewhat later was the investigation by J. Sapin-Jaloustre (1952) during the French expeditions to Terre Adelie Land in the austral summer of 1948–9 and again from January 1950 to January 1951.

The International Geophysical Year (IGY) of 1957 heralded the beginning of the modern era in Antarctica. The Antarctic Treaty was signed and ratified soon after in 1961. This treaty set the continent aside for science, and all signatory nations waived territorial claims for thirty years.

Today transport is often by jet aircraft, expeditions are short (some lasting only two or three weeks), all supplies are brought in, dog sleds are an anachronism, mail arrives at the more accessible stations almost biweekly, and the buildings are large and modern. These changes permit the use of very complex scientific equipment to explore the most technical problems.

There are now so many works recently completed or in progress on the Weddell seal that it would be too much to name them. Several countries have been or now are engaged in such research: Australia, Great Britain, New Zealand, and the United States. The effort has been so great compared to those for other species of marine mammal that we probably know more about the Weddell seal than any other marine mammal.

Why has the effort on the Weddell seal been so intense? The advantages of research on this animal compared to others are many, now that modern transportation has overcome the difficulties of working in Antarctica.

Like animals in many isolated parts of the world, Weddell seals do not respond appropriately, with fear, to man. They do not take flight at man's approach, and are hardly aggressive when disturbed. This phlegmatic characteristic is distinct from the behavior of crabeater and leopard seals, both of which resist intrusion very actively. Such tractability makes the capture and manipulation of Weddell seals relatively easy. They are also abundant in many parts of their range. Within a few kilometers of McMurdo Station, there are between two and three thousand animals in late summer.

Environmental conditions near McMurdo Station could not be more ideal for the study of this animal. The fast ice provides a stable platform for travel over the sea. The conditions are so mild in some places that one could travel in the family car, but in order to cope with all situations four-wheel-drive powerwagons or tracked vehicles are used. In some areas the ice and thin snow cover are so smooth that speeds of 80 km per hour can be achieved with ease. However, there is the risk that where a section of ice was safe one day, a large crack may develop by the next day.

To my knowledge, McMurdo Sound is one of the few places where marine mammals can be collected with wheeled transport such as a truck. One can simply drive up to the animals, net the desired one, place it in a sled, and tow it away. The advantages of this compared to working from a tossing boat are enormous.

The ice also provides a stable foundation on which to set up a complex laboratory over deep water of at least 600 m, in which studies of the seals can be done (Fig. 2.3). I do not believe there is anywhere in the world where oceanic stations can be established with such safety, for such extended periods, and at such low costs. Again the advantages over a tossing ship are great.

Like many animals, Weddell seals congregate at certain localities where they can be seen at almost any time. Unlike most marine mammals, Weddell seals form such congregations where local sea conditions permit underwater observations and practical, meaningful studies can be done. This unique situation is the result of several factors.

Because of the long winter nights and the ice cover, there is very little phytoplankton growing in the water of McMurdo Sound except

Figure 2.3. A study site several kilometers offshore on 2 m thick sea ice and above 600 m deep water. Observation hill in the background is 10 km from the huts. Beneath one hut is a man-made hole for the seal to dive through. The other hut func- tions as living quarters and analytical laboratory. In the foreground is the sled used for moving seals from their normal haul-out areas to the study site. (Aerial photograph taken November 1977.)

for a short time in December. Furthermore, the fast ice of the Sound prevents any surge and stirring of the bottom near shore. Also, because temperatures are below freezing almost all year, there is very little silt-laden runoff from the land. The result is probably the clearest oceanic water in the world. In early spring (October), I have been able to distinguish objects just below the sea ice from over 150 m away. One is also able to see large holes in the ice as diffuse bright spots from as much as 250 m away (see Fig. 1.3 and Fig. 1.4). To my knowledge, such conditions are unequaled in any other place where marine mammals commonly gather.

Finally, because of the Antarctic's past isolation the environment is undisturbed. This means there is the opportunity to study an ecosystem, and the various animals in it, that has not been exploited, contained, or managed by humans, or disturbed by their various kinds of work and pleasure-seeking activities. This situation is virtually impossible to find anywhere else. Unfortunately, it is coming to an end in Antarctica as well.

Suggested reading

Bertram, G. C. L. 1940. The biology of the Weddell and crabeater seals. British Graham Land Expedition 1934–7, Scientific Reports 1:1–39. London: British Museum of Natural History.

Lindsey, A. A. 1937. The Weddell seal in the Bay of Whales, Antarctica. Journal of Mammalogy 18:127–44.

Sapin-Jaloustre, J. 1952, Les phoques de Terre Adélie. Mammalia 16:179–212.

Wilson, E. A. 1907. Weddell's seal, Mammalia. In National Antarctic Expedition 1901–4, Natural History 2:1–66. London: British Museum.

3

Breeding, birth, and growth

Therefore, behold . . . even a marvelous
work and a wonder . . .
Isaiah 29:14.

The sexual behavior of Weddell
seals is a very private matter – unintentionally, at least as far as science is concerned. It takes place underwater, away from the curious
view of biologists.

To my knowledge, only one copulation has been observed. Of all
times, it was during an underwater television taping session by
some University of Minnesota biologists. There is now a permanent
record of the event! Unfortunately, if there was any precopulatory
behavior it was not viewed or taped. This event took place on December 7, 1969. Based on circumstantial evidence, this is near the
end of the breeding season. Microscopic examination of male seals'
testes provide indirect evidence that most spermatogenesis occurs
from mid-September to the end of December.

Although copulation and conception occur mainly in December,
when the seals are most concentrated in and near the pupping areas,
development of the fertilized egg is delayed until mid-January or
mid-February. At this time, the egg implants in the uterine wall and
development goes forth to result in a birth in October.

A prelude to birth and copulation is the establishment by a few
bulls of territories *below* the ice. To my knowledge, this is the only
documented case of a marine mammal holding a territory in the water. Other ice-breeding seals, as well as some whales, may occupy
similar territories, but the necessary detailed studies have not been
done.

Presumably, most of these territories exist near the pupping colonies, and for the months of October through December dominant males spend almost all their time in the water defending their territorial boundaries and excluding other males. The territories vary in size and shape, from about 50 by 15 m to 400 by 50 m (Fig. 3.1). Territorial bulls sometimes share a common breathing hole and probably establish some sort of uneasy truce about this vital place. The bulls are vulnerable when breathing at the surface, and the unwary seal may get his hind flippers nipped. By December, these males start to haul out on the ice. The hazards of their occupation are borne out by the many cuts on their hind flippers, in their axillae, and around their penal orifices.

For the next two months, January and February, the living is easy. The weather is mild, and many animals spend lengthy periods sleeping on the ice. During this time, the old rust brown coat is moulted and a blue-black coat replaces it.

Up to 80 percent of the adult cows are impregnated in the summer and throughout the fall and winter the foetus develops. By early spring a new life and new cycle are ready to begin, and late October sees most of the pups born.

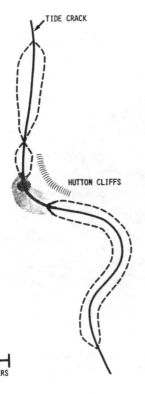

TIDE CRACK

HUTTON CLIFFS

100 METERS

Figure 3.1. Approximate underwater territories of male seals during the pupping and breeding season at the Hutton Cliffs Colony. These boundaries just below the ice surface were defined by haul-out locations and underwater television observation. Depth, the third dimension, is not defined, but seals do restrict themselves to certain boundaries in the water column. (Modified from Kaufman, Siniff, and Reichle, 1975.)

Within about a day after the cow hauls out to have the pup, labor begins. The first signs are rhythmic contractions of the cow's body, accompanied by the raising and lowering of the hind flippers. These contractions last twenty to thirty minutes before the pup begins to emerge. Most births are probably head presentations, but breech births are common (Fig. 3.2). The time from when the membranous sac is first seen until the pup is delivered is less than ten minutes, and sometimes as little as one minute. Breech presentations usually take longer. Shortly after birth, while the mother and pup are maneuvering around each other, the umbilical cord is broken. One to two hours later the cow discharges the placenta.

Soon after birth, both the cow and pup nuzzle and make nose-to-nose contact. During this first period together, there is little vocalization by either animal. By the second hour after birth the pup begins to nurse.

After birth the pup is a wet, 25 kg weakling cast into an extremely cold and windy world. (How it survives this early ordeal is discussed in Chapter 4.)

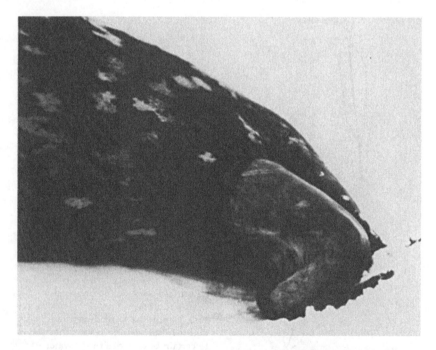

Figure 3.2. A Weddell seal birth in progress. (Courtesy British Antarctic Survey.)

The cow is constantly with the pup for the first week or two. For the next four weeks she takes some leaves of absence, but during all this time the pup is well fed. The mother's milk is initially about 30 percent fat. It becomes progressively richer during the next few weeks until it is over 60 percent fat.

Obtaining this information on milk constituents was something of an adventure as well as a very pleasant experience. It was necessary to milk cows periodically during the six weeks of nursing. In the course of this effort, Chuck Drabek (now a professor at Whitman College but at the time a fellow graduate student at the University of Arizona) and I became well acquainted with some cows and their pups. We worked in one small colony at Turtle Rock, where we would watch for a pup to start nursing. Once it started, one of us would crawl stealthily up from behind and push the pup aside. (The startled look of the pup sometimes made it hard for us to take our job seriously.) Then we would place a length of clear tubing over the active nipple and suck the small sample we required into it (Fig. 3.3).

That is how the procedure worked in theory and sometimes in

Figure 3.3. Charles M. Drabek collecting a milk sample from a cow. The pup in the foreground is nearly ready to wean. (From Kooyman, 1969.)

practice. As the pups got larger it was less easy to push them aside, and sometimes as we were collecting our sample the pup would retaliate with an attack on the top of our parka hood. Such a ruckus would often alert the cow, who would frequently give us a solid slap with the side of her flipper. Then off mother and pup would go, leaving us prostrate in the snow. However, we were successful often enough to determine the fat and water content of milk from cows with pups whose ages were known. In general, there is an increase in fat content with time, and it is inversely and closely correlated with the change in water content illustrated in Figure 3.4.

One of the first to determine what this diet did for the pup was Alton Lindsey, biologist on Admiral Byrd's second antarctic expedition (Lindsey, 1937). Since then, several others have noted similar results. The pup gains about two kilograms per day (Fig. 3.5), and its subcutaneous fat thickens substantially during this period (Fig. 3.6).

During the first two to three weeks, the pup and cow are nearly constant companions; they are separated only when the cow makes occasional short dives. Sometimes the pup will follow the mother to the tide crack, but it usually remains on the ice while she is in the water. If the pup is threatened at this time by another seal, or a person, the mother soon comes out of the water.

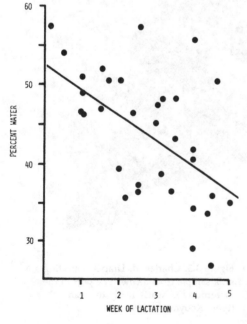

Figure 3.4. Change in water content of milk as lactation progresses. The regression equation is: $Y = 52.6 - 3.25\,x$, where Y = percent water content and x = week of lactation. The correlation coefficient is 0.58. (From Kooyman and Drabek, 1968.)

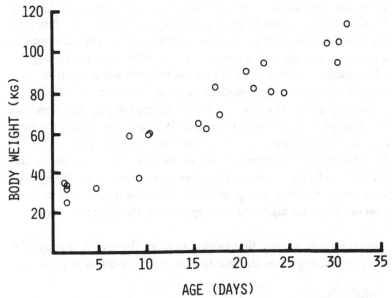

Figure 3.5. Body weight increases of five Weddell seal pups during the first four weeks of nursing. (From Elsner et al., 1977.)

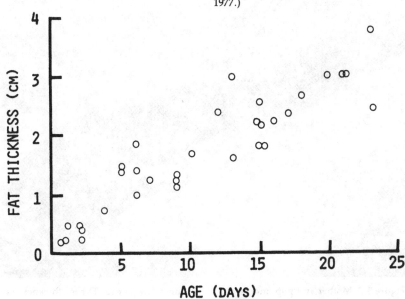

Figure 3.6. The increase in subcutaneous fat thickness of Weddell seal pups during the first twenty-four days of nursing. (From Elsner et. al., 1977).

When the two animals are separated, there is much calling by each. At reunion, there is nose touching and exploring of bodies with the snout. The final or conclusive means of recognition seems to be smell. In addition to the impressions gained from observing the usual reunion of animals, it has been noted that when a pup is contaminated by the smell of another pup the cow seems confused. In one case of this sort, the cow refused to accept the pup and deserted it. If a strange pup approaches a cow, it is usually driven off by a threatening snap or even an actual bite. Adoption is rare.

Considering the tremendous drain on the fat resources of the mother, the selective pressures opposing adoption must be strong. When a mother allows a pup other than her own to nurse, neither pup seems to get enough and their growth is much slower than normal.

When the cautious pup first begins to enter the water, it is usually after much coaxing from the mother. Sometimes the cow resorts to

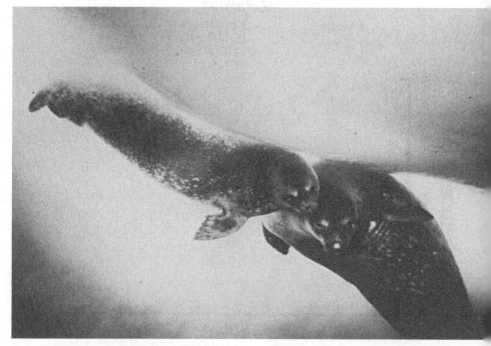

Figure 3.7. Mother and pup underwater at a breathing hole near Turtle Rock. The mother is very protective of her pup and may threaten other seals if they approach within a few meters. The ice thickness is a few centimeters at the center of the crack and rapidly increases to 2 m a short distance away.

force and drags the pup in. However, she will also help the pup to get out if it has trouble doing so. This is usually done by floating up under it and providing support or by simply giving it a push.

In the water, the cow stays close to the pup (Fig. 3.7). They often do much maneuvering around each other that is suggestive of play. During these periods in the water, the female is as protective of the pup as she is out of the water. If a bull seal approaches within about three meters, she gives chase. Similar threats, snaps, and nips are imposed upon any cows that approach too close to the pup. Such encounters are sometimes by accident when the pup becomes confused and swims toward the wrong female.

Within fifty days, the cow deserts the pup. At this time, the pup may weigh well over 100 kg. But its gain has been the cow's loss. As a result of feeding the pup and fasting, she may lose from 100 to 150 kg. From a round robust figure in October, the cow becomes by December but a shadow of herself (Fig. 3.8). Through the winter, she will fatten up once again and be prepared to start the cycle over again in the fall.

After weaning, the pup loses weight while it develops its diving skills. A pup only six to seven weeks old can already remain submerged for at least five minutes and dive to depths of nearly 100 m.

For the next three years, most of the young seals seem to disappear from the McMurdo Sound area. Little is known about the seals during this period. A few pups remain in the area, and some, if not most, of these animals do poorly. Two pups that we brought back to Scripps in 1968 probably grew at about the rate of a successful, wild pup; within about a year each weighed 200 kg.

There is virtually no information on the social behavior of these young animals. One year in late summer, I found 10 to 15 pups collected together near Cape Armitage. These pups seemed to be traveling together. Also in late summer, during a helicopter flight to Dunlop Island (which is 75 km northwest of McMurdo Station), I saw a collection of 150 to 200 seals at a large pool next to a sizable iceberg that was held fast in the sea ice. I estimated that 70 percent of these seals were pups, yearlings, or subadults. Most interesting to me was that, for the first time, I saw Weddell seals "playing" or sparring. This occurred in a large pool next to the berg. The only other time I have seen this kind of "play" was with our captive animals, which spent long periods of time chasing each other around their pool.

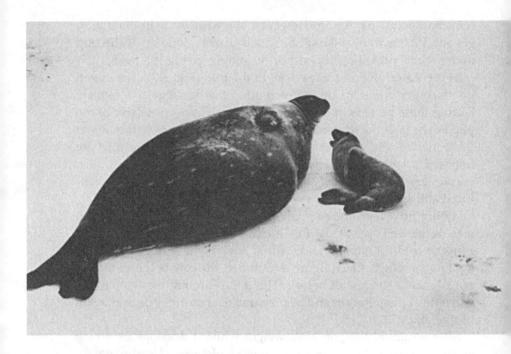

Figure 3.8. (*Above*) Mother and pup in October, shortly after the pup's birth, when it weighs about 25 kg and she weighs about 450 kg. (*Left*) Mother and pup in December, shortly before weaning. At this time, the pup (foreground) weighs about 100 kg and the mother weighs about 300 kg. (Photos by M.A. Castellini.)

When three years old, some of the females may breed and in subsequent years join the pupping colonies in early spring. It is not known at what age the males assert themselves effectively enough to control territories.

Suggested reading

Kaufman, G. W., D. B. Siniff, and R. Reichle. 1975. Colony behavior of Weddell seals, *Leptonychotes weddelli*, at Hutton Cliffs, Antarctica. *Rapports Procès Verbaux Réunions, Conseil International Exploration Mer.* 169:228–46.
Siniff, D. B., D. P. DeMaster, R. J. Hofman, and L. L. Eberhardt. 1977. An analysis of the dynamics of a Weddell seal population. *Ecological Monographs* 47:319–35.

4

Cold

We dwelt on the fringe of an unspanned
continent, where the chill breath of a
vast, polar wilderness, quickening to the
rushing might of eternal blizzards,
surged to the northern seas. We had
discovered an accursed country. We had
found the home of the blizzard.
Sir Douglas Mawson, The Home of the
Blizzard

This chapter follows the title of a
book written some years ago by Laurence M. Gould, Chief Scientist
on Admiral Richard E. Byrd's first expedition to the Antarctic. It is
the most succinct way of describing the antarctic environment and
perhaps the major environmental factor with which Weddell seals
must cope. Under all circumstances they must maintain their body
temperature at 37°C. They must do this while living in a region
where the water is constant at −1.9°C. The air temperatures range
from 0° to −50°C, and concomitant winds may create chill factors of
−100° to −150°C.

The atmospheric conditions are usually not significant to the adult
because, if the situation becomes too uncomfortable, the seal can always return to that great ameliorating agent: water. However, for the
first week of life this may not be possible for the pup, and I will
discuss shortly how it maintains its body temperature in a way different from that of the adult. Before discussing any of the ways in
which the seals, adults or pups, regulate their body temperature, I
would like to review briefly some physical characteristics of heat
transfer that might help in the following discussions.

Whenever objects are at different temperatures, such as the warm
Weddell seal and the surrounding air, ice, and water, heat will flow
by means of *radiation, conduction* and *convection*, and *evaporation*.

Radiation is that exchange of electromagnetic energy between two
objects which is dependent only upon the temperature and nature of
the surfaces of the radiating objects. The sun, with its surface tem-

perature of 5,760°K (6,033°C), is radiating a considerable amount of energy. How much reaches the seal depends upon a variety of atmospheric conditions as well as solar angle. During a clear summer day at peak sun angle, solar radiation flux may be close to 1,000 watts per square meter (Kerslake, 1972). The dark seal in turn absorbs this radiation nearly as well as a blackbody within the wavelength of 0.5 to 1.0 μm, or the peak intensity of the sun's visible radiation.

In some cases, if the temperature difference between two surfaces is not great ($\sim$ 20°C), an approximation of the Stefan-Boltzmann law of radiation is satisfactory. The rate of heat transfer in watts per square meter is dependent upon the temperature difference to the first power. In this case, $\dot{Q}_R = C_r (T_1 - T_2)$, where C_r is the radiation coefficient for heat exchange in watts per square meter per degree centigrade and T_1 and T_2 are the temperatures of the two surfaces. In short, the rate of heat transfer by radiation ($\dot{Q}_R$) from the seal to the environment is directly dependent on the seal surface temperature and environmental temperatures.

Conduction is the direct physical transfer of thermal energy through liquids, solids, or gases. The amount of heat transfer is dependent upon the path length, nature of the material, the thermal gradient, the area, and time. Thus, $\dot{Q}_c = k \cdot A (T_2 - T_1)/\ell$, where $\dot{Q}_c$ is the rate of heat transfer expressed in watts, k is the thermal conductivity of the medium watts per meter per degree centigrade, A is the surface area of the conducting medium, T_2 and T_1 are the temperatures in °C of the surfaces, and ℓ is the separation distance in meters between the two temperatures. For example, the surface temperature of the seal's skin may be 2°C, the muscle temperature 5 cm below the skin may be 35°C, and there may be a steep temperature gradient through the intervening blubber. Heat loss from the muscle to the skin is low because blubber is a good insulator or poor conductor; its conductivity is about equal to that of asbestos. If there is blood circulating from the muscle, then heat loss through the blubber to the skin is greater as the warm blood from muscle carries heat to the skin surface. In other words, when there is a fluid (blood) or gas medium between two objects (i.e., skin and muscle) then heat transfer is enhanced if there is mass movement of the medium, which acts as a physical carrier of heat. This process is called *convection*.

Finally, during the process of *evaporation*, a considerable amount of heat can also be transferred. For example, 676 cal (2,830 joules) are required to convert ice at 0°C to water vapor. Thus, loss of heat by evaporation depends upon the change of water to vapor and the rate at which it diffuses. The evaporation equation is expressed as $E = C_e(P_1 - P_2)$, where C_e is the vaporization coefficient expressed as 0.133 watts per square meter per millimeter of mercury, and P_1 and P_2 are the water vapor pressures at the skin surface and of the ambient air.

With these physical principles in mind, let us now consider the question perhaps most frequently asked by both lay people and scientists about mammals that live in the polar regions – how do they keep warm? Perhaps the best place to begin is at birth, when the pup is cast out from the warm mother's womb into the harsh antarctic spring weather.

The pup is born wet (Fig. 4.1), but probably little heat is lost by evaporation because moisture on the fur rapidly freezes and falls away in the form of ice crystals. (Indeed, as water freezes some small

Figure 4.1. A Weddell seal pup moments after it is free of the amniotic sac. (Courtesy of British Antarctic Survey.)

amount of heat is released (80 calories per gram of water to ice).

The neonate pelt is strikingly different from that of the adult. It is a long, thick fur similar to that of many terrestrial fur-bearing mammals, and in a thermoregulatory sense it is more similar to a fur-bearing mammal (Fig. 4.2). This thick fur, frequently called lanugo, will last for only a few weeks. The shedding begins 9 to 21 days after birth and is complete by 44 days. During this time, the lanugo is gradually replaced by the shorter, more sparse hair of the adult. By this time it has served its purpose of providing insulation to the pup, who in the interim has accumulated a thick layer of insulating blubber (which accounts for most of the great weight gain mentioned in Chapter 3; Fig. 3.5).

The evidence of a team of field investigators, headed by D. D. Hammond, that studied the early thermoregulatory development of the pup shows that it is essential that the lanugo's role is eventually

Figure 4.2. A newborn Weddell seal pup and its mother. The pup has a luxuriant coat of long fur (lanugo), which will be replaced in 4 to 6 weeks by short fur like that of the adult.

superseded by that of blubber. This fur is not waterproof, and if the
pup should enter water too soon in life it risks chilling. For example,
an 8 hr. old Weddell seal pup was placed in cold seawater for 10 min.
During this time, its body temperature rose from 38° to 39°C as much
energy was expended and heat produced while the pup thrashed
about. It was then removed from the water and for the next 15 min.
its body temperature declined to nearly 36°C (Fig. 4.3). This was because the wetted pelt provided no insulation and there was much
heat lost through radiation, conduction-convection, and possibly
even evaporation. Nine days later, after some subcutaneous fat had
accumulated, this pup was given a similar dunking with no effect on
body temperature. The transition had been made, and the pup was
now a marine mammal in a thermoregulatory sense. It possessed at
this time a layer of blubber, a material of low conductance.

    Similar findings have been obtained by Soviet scientists who stud-
ied the newborn Greenland or harp seal, *Histriophoca groenlandicus*,
(Davydov and Makarova, 1964). They noted that when newborns (1

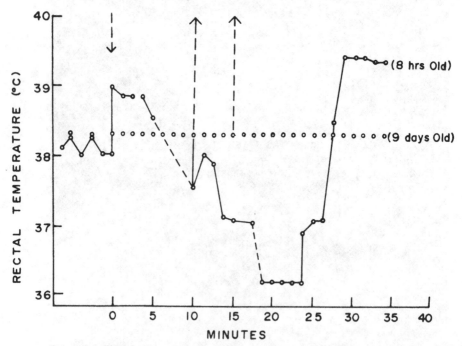

Figure 4.3. The rectal temperatures of a   at −1.9° C. The arrows indicate when im-
pup when 8 hours and 9 days old, before,   mersion began and ended. (From Elsner
during and after immersion in seawater   et al., 1977.)

to 7 days old) were wetted in seawater for 30 min. the metabolic rate was more than twice that before wetting, but in the 30-to-40-day-old animals no increase in metabolism occurred.

It has been noted by some investigators that neonate seals have a higher metabolic rate than is usual for their size (Grav, Blix, and Pasche, 1974). For the first few hours of life, there could be some distinct advantage to a high heat production in order to compensate for high heat conductance while the pup is drying out and overcoming the initial chill following birth. Intriguingly, studies of harp seals may have uncovered a clue to the source of energy for this high heat production. Unusual adipose tissue has been found in the newborn pups. This localized tissue is just below the skin in the nucal and sacral areas, where it is, respectively, 2 to 8 mm thick. Electron microscopic examination shows it to have a structure similar to the brown fat structure of other animals. The exciting aspect of this discovery is based on the known function of this tissue in small terrestrial mammals. In these animals, it is an important high-energy fuel. For example, in those animals that hibernate it makes possible a fast warm-up and rapid arousal from the hibernal state. It does this by a unique process of consuming large amounts of oxygen, which is converted into heat (and into little of anything else, such as mechanical energy). Consequently, such a process could be important in the newborn seal by turning it into a nonshivering heat generator to compensate for the high heat loss of the wet, poorly insulating pelt.

After weeks of nursing, accumulation of blubber, and loss of lanugo, the pup has adopted the same modes of heat conservation as the adult. A thick blanket of blubber surrounds the entire animal except the extreme portions of the flippers, snout, and eyes (Fig. 4.4). This mode of insulation is present in not only the Weddell seal, but

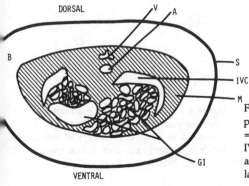

Figure 4.4. Cross-section of a grey seal posterior to the chest; A = dorsal aorta, B = blubber, GI = gastrointestinal tract, IVC = inferior vena cava, M = muscle, and V = vertebra. (Adapted from Scholander et al., 1950.)

all other seals, the sea lion, the walrus, the sea cow, and all whales. The only marine mammals that rely on fur for insulation are the fur seal and sea otter.

Considering the attributes of the two modes of insulation, one wonders why fur as an important insulator has been retained at all in marine mammals. It requires a considerable amount of care and maintenance. Both fur seals and sea otters spend 10 to 20 percent of their time grooming. Also, fur-bearing animals seem to have less tolerance to high temperatures. If driven or herded on land, they quickly become overheated. On warm days, fur seals soon resort to flipper fanning and shortly resort to moving into the water. Presumably, this is because they cannot radiate or conduct heat well through the fur, and their main outlet is through the limited surfaces of the flippers.

By comparison, blubber is an active tissue with a good, controllable blood supply. During cold exposure, little blood flows through the blubber and to the skin. A steep temperature gradient exists between the muscles and the water or air (Fig. 4.5). At this time, the conductivity coefficient of blubber is about 0.0029 watts per centimeter per degree centigrade, or approximately the same as asbestos. On the other hand, if the activity of the seal, or the environmental conditions, dictate that it must lose heat, hot blood is carried to the surface and much of its heat radiated and conducted away, not only from the flippers as in fur seals, but also through the general body surface area allowing the entire body to play an important role in

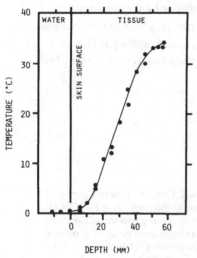

Figure 4.5. The temperature gradient from the skin to the muscle of a harp seal immersed in icewater. The first 40 mm of tissue is probably blubber. (From Irving and Hart, 1957.)

heat dissipation. The sparse hair of seals and sea lions presents little barrier to this heat loss, and sea cows and whales have no hair at all. This ability to lose heat is so effective that we have noted a core temperature drop of 2° or 3°C in harbor seals within 2 or 3 min. after the administration of vasodilating drugs.

Furthermore, the body mass is buoyed by water, the extra bulk of blubber presents no problem, and locomotion in water is exceptionally graceful. For those groups that come on land (seals and sea lions), the extra mass is another matter. Movement is slow and awkward, but time spent ashore is usually for resting or to nurse pups and their terrestrial clumsiness is seldom a liability.

Blubber also behaves like a coat of armor, protecting sensitive body parts from bruises and cuts. And, finally, it is a built-in energy store that is available: (1) when food sources are scarce; (2) when prolonged fasts are necessary during migrations; and (3) when animals must remain ashore for extended periods, such as in the breeding season, when male sea lions and fur seals control territories and must be present at all times.

As good an insulator as blubber is, other compensations are apparently necessary for existence in the medium of water, which, when still, has a heat conductivity coefficient 25 times that of air and probably more than 100 times that when the seals are swimming. For most marine mammals of which the metabolic rate has been determined, it has been found that its metabolism is higher than similarly sized terrestrial mammals. The values for seals presented in Figure 4.6 are based on total body weight. The consistency of these high metabolic rates in marine mammals seems to suggest that greater heat production is generally necessary in these forms to overcome the greater heat losses to the environment, even though there are several exquisite ways of controlling heat loss.

To conclude, the sea is a cold environment for an endotherm; even in the warmest tropical seas the temperature difference between the seal's or whale's central core and the water is 10° to 15°C, and heat is always flowing from the core to the water. In the Antarctic, this is especially true, and heat conservation is necessary with few exceptions. An example of one exception might best aid in summarizing this chapter. On cold days, a seal on the ice surface conserves heat by reducing blood flow to its flippers and skin. which chill to as low as a degree or two above freezing. Under these circumstances, little heat is lost to the surface through radiation and conduction, and new-

fallen snow does not melt from its fur (Fig. 4.7). On rare occasions
when there is no wind and the sun shines brightly, the seal is subject
to a heat burden and must lose heat even while quietly sleeping. At
this time, one may see the incongruous sight of a seal steaming as
moisture evaporates from its body and flippers. The seal will also
wave its expanded flippers in the air to increase conductive and con-
vective cooling. In the seal's shadow, more heat is lost through radiation from its body. Heat is conducted from its hot skin to the snow
and ice on which the seal rests. This leaves a melted pool of ice water,
which soon freezes after the seal departs. Remaining is a snowcast, a
snow relief record of the seal's presence (Fig. 4.8).

Suggested reading

Schmidt-Nielsen, K. 1975. Temperature regulation. In *Animal Physiology: Adaptation
and Environment*, pp. 296–368. Cambridge: Cambridge University Press.

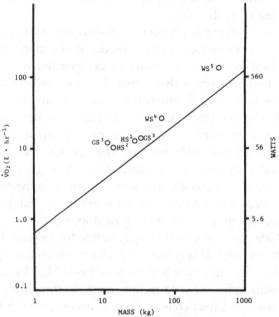

Figure 4.6. The standard regression plot
for terrestrial mammals where $\dot{V}_{O_2} = 650$
$M_b{}^{0.75}$, expressed as $mlO_2 \cdot hr^{-1}$, and
M_b = kg of body mass. HS = harbor seal,
GS = Greenland seal, and WS = Weddell
seal. The superscript numbers refer to the
references in Table 7.1.

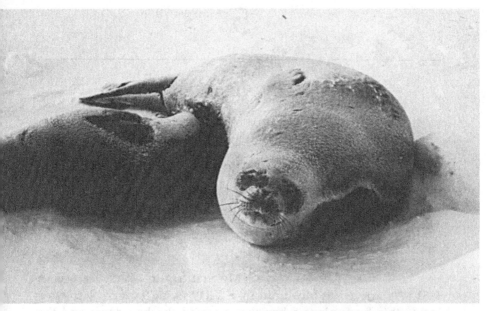

Figure 4.7. A Weddell seal mother and pup. Both the mother and pup have layers of new-fallen snow on their coats that remain unmelted except around the eyes.

Figure 4.8. Two imprints where Weddell seals had slept and melted the snow. The water refreezes to form snowcasts.

5

Diving behavior
POSEIDON'S PRIDE

GLENDOWER: I can call spirits from the
vasty deep.
HOTSPUR: Why, so can I, or so can any
man;
But will they come when you do call for
them?
Shakespeare, King Henry IV, Part I

My first trip to Antarctica was in
1961. We arrived at McMurdo in a C-119 flying boxcar. It snowed
inside this huge plane when they popped the bay doors and the
frigid outside air cooled the warm moist air inside.

McMurdo Station reminded me of a frontier-town movie set. It
was a hodgepodge of prefabricated plywood buildings and canvas
Jamesways (Fig. 5.1), which are half tubes with a doorway at each
end. No one had regular hours of sleeping, nor did they spend much
time doing so. The twenty four hours of light were responsible for
the unscheduled sleeping patterns, and there was little incentive to
remain in the bunkrooms any longer than necessary.

Our sleeping quarters were a long canvas Jamesway that was dark at
all hours. It was heated with one fuel oil stove that had no fan so that
severe temperature gradients existed. Those in the top bunks were
roasting in dry, desertlike heat while those in the bottom bunks
were bundled in down sleeping bags for comfort against the below-
freezing temperatures. The nearest washroom was 200 to 300 m
away.

All these inconveniences were overshadowed by the excitement of
being on both a geographical and scientific frontier. Such a spirit has
been dulled somewhat with the modern buildings, television, and
soft drink dispensers now at the stations, along with the more re-
fined methods of subsistence and transport of people.

Every opportunity was taken to be away from the base and it was
on one of these occasions when I first saw a Weddell seal. Even then I

knew it had great scientific potential. This did not require much keen insight on my part. The seal was lying next to a small hole it had made a short distance offshore. To reach the seal, Art DeVries, John Dearborn, a specialist in benthic polar organisms, and I had driven a truck out over the two meter thick sea ice. When we arrived, the seal paid no heed to us. Two characteristics of the Weddell seal (described earlier in Chapter 2) are among the most remarkable attributes associated with a marine mammal: (1) The seals emerge from isolated holes in thick, stable ice, where it is possible for large vehicles to travel; and (2) they are indifferent to humans. What was equally mind boggling was that no one had up to this time taken advantage of these attributes in the studies of the Weddell seal.

One year later, I was enrolled in graduate studies in the Zoology Department at the University of Arizona and was diligently trying to instill my major professor, William S. MacCauley, and the department head, Albert R. Mead, with some of my enthusiasm for a thesis project on the Weddell seal. For men who had until that time never

Figure 5.1. Mainstreet McMurdo Station.
(Photo taken February 1962.)

heard of Weddell seals, they showed a lot of patience and open-mindedness. The following year I went back to McMurdo Sound.

In the beginning, I was not at all sure of the best approach. I knew that individual animals must be identifiable, and I had flipper tags to help confirm identifications derived from various facial and body markings. Also, I wanted to attach and recover depth recorders at will. The previous year, DeVries had put depth recorders on several seals, but recovery had been fortuitous because he had relied on finding the seals hauled out two or three days later. The measurements he obtained were useful in the design of my recorders because they provided information on the pressures that the instruments would have to withstand.

The method of study became firmly established as a result of another investigator's problem. Donald "Curly" Wohlschlag, then a professor of ecology at Stanford University and my boss on the first trip to Antarctica, needed to catch some midwater fish. To do this, he planned to guide remotely a diver's motor sled between two ice holes about 100 m apart. Attached to the sled would be a line that could then be used to draw the net between the holes – perhaps a logical idea, but technically difficult. What happened was that the sled guided the line to the bottom instead, and the retrieved unit was a crinkled wreck.

DeVries then came up with the novel idea of hooking the line to a seal and letting it swim between holes. This may have been even before the U.S. Navy began using marine mammals in their missile and torpedo recovery program. There was consternation that what might happen would be an incredibly tangled line and a drowned seal. As a navigational aid, a small, 60 kg seal was placed in one hole; fortunately it remained there. Then a 200 kg seal was captured with much effort. In fact, the release scene was mayhem, and the seal was only too glad to leave with the line, one of my depth recorders, and almost a couple of biologists. When the seal departed, all of us smirked a bit wondering how a phlegmatic biologist named John MacDonald, who was waiting at the other hole, would fare taking the line from this very irritated seal. Apparently, the swim cooled the seal off; John survived to become a faculty member of a New Zealand university. Wohlschlag strung his net between two ice holes, and I had my experimental procedures. Soon thereafter, I was blasting holes well offshore and capturing and moving seals to them from various parts of the Sound.

The first hole was at a site one mile offshore from Cape Armitage. The ice thickness here was 4 m, and it took two weeks of back-breaking labor to make the hole. My tools were a chain saw, ice tongs, and, for the last few feet, explosives. Peter Koerwitz, the biology laboratory technician, helped with what was an impossible job for one man. Now they use a 1.2 m powered augur, which cuts the holes in one to two hours.

Once the hole was cut, a heated hut was placed over it. This prevented the hole from freezing over and also provided a comfortable laboratory in which to work. Water depth below the hole was 300 m, and it was an ideal oceanographic station without the inconveniences of shipboard life – the constant noise of generators, rolling and pitching movements, restriction to the ship until it returned to port, the daily fuel and crew expenses, and many others.

Once the laboratory was set up, a seal was netted, placed in a sled, and towed to the laboratory with one of the various tracked vehicles at the base. At the hut, the seal was made ready by anchoring one or two key rings to its hide with hog rings. When I first considered this method, it seemed a bit repulsive, but after some thought and experience it seemed a sensible approach. It was fast, which is important when you are dancing around inside a 1.2 by 3.7 m sled with a 450 kg seal; it was not very painful, which is also important when you are sharing such a small sled with an already irritated animal. More important, this method offered much less resistance to swimming than a harness, and if the seal were to get away and we failed to find it the hog rings would pull out in a short time, whereas a harness would not. With the key ring as a place for attachment, it was an easy matter to put on and remove small packages from the seal while it was in the water.

When I first conceived this project, the major obstacle was obtaining small, cheap depth recorders. None were commercially available. Then I read a report by P. F. Scholander (Scholander, 1940), an eminent respiratory physiologist and director of the laboratory that I later joined at Scripps. He described his method of measuring the depth to which harpooned fin whales were diving. It was a long, glass capillary tube, closed at one end, the interior of which was dusted with a water-soluble dye. The maximum depth was recorded by leaving a ring at the point of maximum compression inside the tube. By measuring this distance, it was a simple calculation to obtain the maximum depth to which the tube had been exposed. (I should mention that this device was not original with Scholander. In the 1850s, Lord Kelvin constructed a somewhat similar recorder for use by ships as a depth-sounding device. Previous to this, ships used a drop line that required them to stop each time they made a sounding, a great inconvenience for sailing vessels. Thus, Kelvin's recorder was a major innovation because it made soundings possible while underway. A hundred years later, the same principle is being used to produce the cheapest and most popular depth gauge for scuba divers.)

Soon after my discovery of Scholander's report, I was acquiring all the glass capillary tubing at the University of Arizona. Sixty centimeters was the estimated appropriate length to give me the resolution I needed on the deeper dives. The glass was bent into lengths of about twelve centimeters and wrapped in neoprene. In this configuration, they could be put on and taken off seals with ease. However, these devices gave only maximum depth and were good for only one or two dives.

In my quest for a small time-depth recorder (TDR), I searched in vain through catalogues and scientific reports. All those available were too large to suit my needs. This was discouraging because such an instrument was essential for a detailed study of diving behavior. It would provide a host of otherwise impossible-to-obtain information, such as rates of descent and ascent, frequency of time at any given depth, an overall profile of depth against time for all dives, and much more. Most importantly, it was essential for recording the duration of dives. There is no other way of confirming whether a seal had surfaced or not. This was to prove especially important when the then almost unbelievable submergence times were determined.

There was no other course of action than to design and build custom recorders. To do this, I cajoled Howard Baldwin of Sensory Systems Laboratory into helping me. His interests were electronic and he was not especially excited about the mechanical recorders I had conceived, but with his help and that of a local jeweler, Bernard Strothman, six instruments were made that weighed 0.5 kg out of water, were 7.6 cm in diameter and 7.6 cm long, and worked admirably well (Fig. 5.2). They consisted of bourdon tubes removed from special high-vibration-tolerant gauges, glass discs greased and dusted with fine charcoal, and one-hour kitchen timers. The assembled parts were mounted in machined brass cans. Later models would run longer, but the basic design remained the same.

The first seal I brought to the hole was a small animal of 150 kg. A TDR was put on just before the seal went into the water. It made one short dive of 2 or 3 min. returned for a few breaths, and disappeared. A deep concern prevailed. Had I drowned my first experiment? By a stroke of luck, a few hours later I learned what had happened to Number 1. Murray Smith, a New Zealand biologist studying Weddell seal population dynamics, was counting seals near Cape Armi-

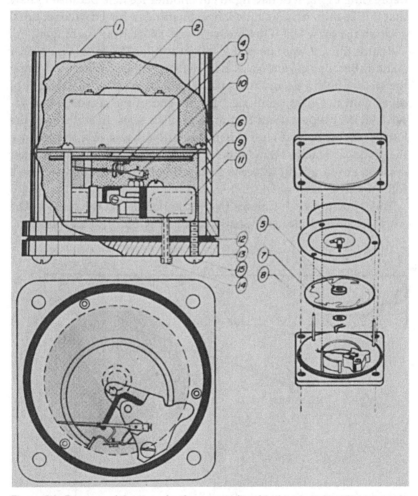

Figure 5.2. Diagram of the time-depth recorder: (1) casing; (2) timer housing; (3) timer shaft; (4) glass disc bushing; (5) set pin; (6) smoked glass disc; (7) washer; (8) locking clip; (9) brass timer mounting stud and locking screws; (10) pressure indicator arm and needle; (11) bourdon tube; (12) "O" ring; (13) brass cover; (14) bourdon tube external opening; (15) locking screws. (From Kooyman, 1965.)

tage that afternoon. While doing so, Number 1 popped up with the TDR pack on its back. The surprised Smith removed it and returned it to me. The recorded profile showed that the seal had not gone deeper than 55 m, and had apparently swum nearly directly back to Cape Armitage (Fig. 5.3). The 1.5 km trip had taken 26 min. This result showed clearly that the seal was too close to shore and the hut would have to be moved further offshore. I was buoyed up from this depressing thought of having to cut another ice hole because I knew that this 26 min. dive was the longest natural or forced dive reported at this time for a seal. Obviously, it was an exciting harbinger.

By this time, it was late in the season and too late to set up another station. Because several seals had discovered and were now using the station hole, I spent the rest of the time recording their depth of dives with the glass capillaries. By the end of the season, I had records of the deepest dives ever measured for seals, in addition to the longest dives. Armed with this information, I was able to convince the National Science Foundation to continue my support for another season. I even got an increase in funding. It was enough to hire a full-time field assistant.

The following year, Charles Drabek joined me on the project. The work progressed rapidly because of the previous year's experience.

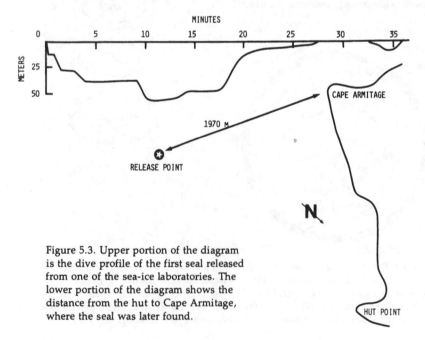

Figure 5.3. Upper portion of the diagram is the dive profile of the first seal released from one of the sea-ice laboratories. The lower portion of the diagram shows the distance from the hut to Cape Armitage, where the seal was later found.

We set up our station 10 km offshore, and at this distance the seals did not escape easily to other breathing holes. Occasionally, we had to move because leads, or wide cracks, would appear nearby that permitted the seals to reach the surface. Moves were not as painful, though, because the ice was only 2 m thick. Also, Thomas Poulter, the scientific leader of Admiral Byrd's Second Antarctic Expedition and an expert in seismic studies, was visiting McMurdo Station. He shared some of his knowledge of explosives. Poulter tutored us in the proper method of blowing ice holes, which eased our task considerably.

At our most permanent station, we placed an under-ice observation chamber 10 m from the breathing hole. A considerable amount of information was learned from many hours of direct observation, records of maximum depth, and depth–time recordings. Those hours in the hut and observation chamber were some of the most peaceful of my life.

Adult seals were the easiest to study because they were less concerned about activities going on above them. This permitted easy attachment and removal of instruments. Often an adult would become territorial about the hole and attack other seals released into it. This resulted in some spectacular fights and our underwater post provided us with ringside seats.

One adult male became so attached to the hole that he remained there even after he had found another breathing hole. For nearly a month of our studies he would haul out and rest at the other hole and then come to ours to hunt. Perhaps the incentive was his fishing success at the station. He caught several *D. mawsoni*, all of which weighed 25 kg or more (Fig. 1.5).

When a seal was first released, it would usually make a series of short, shallow dives to familiarize itself with the surroundings. These dives were most often 5 to 15 min. in length and no deeper than 200 m. After this, exploratory dives would begin to occur from time to time. (Such dives will be discussed at length in Chapter 8.) They are one of the most interesting dives the seal makes, and because of the nature of the conditions the seals probably made more of these dives during our experiments than normal. These dives are always relatively shallow; the average maximum depth was 130 m, few were recorded to depths greater than 200 m, and most of the dive period was much shallower. The minimum dive duration was 20 min., and the maximum certain recording was 73 min.

At one time, we had a seal swimming to and from a natural breathing hole 2 km away from our station while a TDR was attached. Assuming that it was a direct trip, which the dive profile indicated, the animal swam at about 10 to 12 km/hr. Based on these estimates, the seals, during extended exploratory dives, may make round trips of 10 or 12 km, and occasionally one-way trips of equal distance.

Our best example of an animal that may have extended itself this far came in a rather unexpected manner a few years after my original study. Robert Elsner, a cardiovascular physiologist interested in asphyxial defense responses of diving mammals, and I were collaborating on a project on Weddell seals. He had proposed that pregnant seals might not be able to dive as long because the foetus would be parasitizing the mother's blood oxygen stores (more on this in Chapter 6). Therefore, I was diving pregnant, near-term animals to see if this was so. The study was inconclusive because motivation plays a significant role in a seal's willingness to make an extended dive. Clearly, these animals were highly motivated. They were in the last days before pupping and thus suddenly to find themselves in water, a long way from the traditional pupping localities, and unable to get out must have been disturbing to them. Among these females I never before observed such a concentrated effort of prolonged diving.

One of these seals did not return on her second dive after being released. A short time later, she was found on the ice at a rookery 15 to 20 km from the station. Analysis of the TD record showed that after 50 min. of submersion she had reached a depth of 70 m and was still descending (Fig. 8.5). Unfortunately, the timer stopped at this point and the total length of the dive is unknown. This and other almost as dramatic examples have left me with the impression that there are few places the seals cannot reach in the Sound if so inclined.

After several hours, if the seals were unable to find another exit they would usually begin another diving routine. These were dives of a consistent duration of 8 to 15 min. and surface times of 2 to 4 min. This kind of activity sometimes went on for 8 or 9 hr. without a break in the routine (see Fig. 5.5). These dives were easily recognized from underwater observations because the seal would leave the hole by swimming vigorously away at an angle of 75° to 90° from horizontal. On their return, they ascended even faster. The momentum from the deeper portion of the ascent allowed them to coast at least the last 50 to 75 m up into the hole. Their return was also from an equally steep angle. The average rate of descent for 33 measurements was 35 m per min., and the maximum ever measured was 115 m per min. The average rate of ascent was 50 m per min. and the maximum was 120 m per min. The depth of their dives ranged between 200 and 400 m. Because of their regularity, their quantity, and when they occurred during the experiments, I am convinced these were hunting dives.

Although most dives were less than 400 m, there were a few that were more. In order to try to determine the maximum depth to which the seals in the Sound would dive, a new kind of depth recorder was used. This was a commercial recorder that scribed a line on a glass slide coated with gold foil. It could be left on for many hours and still record accurately. The first unit employed had a maximum depth limit of 500 m. It worked well.

After one series of dives by a large male, I was examining the trace while eating a dry cracker. The trace caused me to choke on the cracker. It exceeded the recorder's maximum by a substantial amount. The scribe had returned to zero, indicating that the sensing element was not damaged, so a series of test drops were made to determine the actual depth. A duplication required a drop to 600 m, the deepest dive ever recorded up to that time for an untrained, unharpooned marine mammal. However, carcasses of sperm whales have been found entangled in deep sea cables at greater depths, and presumably these drownings occurred while the whales were hunting (Heezen, 1957). Recently, several sperm whales have been tracked with echo-ranging equipment to depths as great as 1,140 m (Lockyer, 1977). Over the years, using depth recorders with 1,000 m limits, several dives near 600 m have been recorded for Weddell seals. Clearly, from the number they are rare.

With this knowledge of the Weddell seal's capability for deep and prolonged dives, it was logical to pursue studies that might better define the structural and functional basis for this ability. After completing my thesis work, I had the good fortune to spend a year in the laboratory of Richard Harrison, then at the London Hospital Medical School. He has contributed more extensively to our knowledge of marine mammal anatomy than any other investigator. Following that year, I moved to Scripps Institution of Oceanography, where I began and am still conducting physiological studies that form the basis of the following chapter.

Over the past few years at Scripps, a new type of TDR has been designed and built (Fig. 5.4). This unit normally records every dive for two and a half weeks, but because of the intense antarctic cold the batteries fail after about nine days. Even so, this increased the capabilities of monitoring under-ice activities enormously. It has also made it possible to conduct studies on free-ranging animals (those without the constraint of an isolated hole and free to go anywhere above and below the ice). The record presented (Fig. 5.5) is the dive schedule of one such animal. As you can see, the effort is considerable and relentless. From the jagged nature at the bottom of the dive, the seal is moving up and down, probably as part of a very active pursuit of fish. We do not know from these records how far from the sea floor the seal is. In some instances, the depth of dive is so consistent that it suggests the animal is working close to the floor of the Sound.

Suggested reading

Kooyman, G. L. 1968. An analysis of some behavioral and physiological characteristics related to diving in the Weddell seal. *Antarctic Research Series* 11: 227–61. Biology of the Antarctic Seals III, ed. W. L. Schmitt and G. A. Llano, Washington, D. C.: American Geophysical Union. 1969. The Weddell seal. *Scientific American* 221: 100–16.

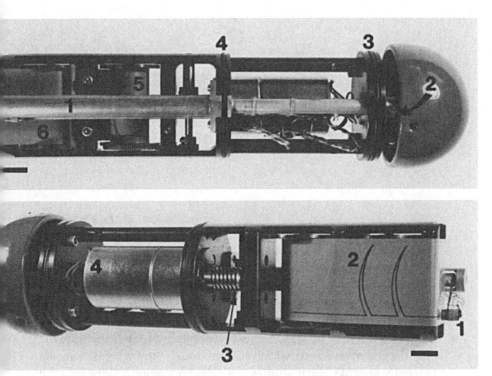

Figure 5.4. (*Above*) Top view of the time-depth recorder: (1) housing for the pressure transducer; (2) outside part of the pressure transducer; (3) "O" -ring seal of the housing; (4) "O" -ring light seal for the film; (5) film takeup spool, (6) film supply spool. Two AA-size lithium batteries fit in the space between the two "O" -ring bulkheads. The solid bar equals 1 cm. (*Below*) Bottom view of the TDR. (1) plotting arm with a light-emitting diode mounted at the tip; (2) the film with a mock record drawn with ink; (3) gear train to drive the film takeup spool; (4) 12-volt electric motor. The solid bar equals 1 cm. (Photos by J. O. Billups.)

WEDDELL SEAL

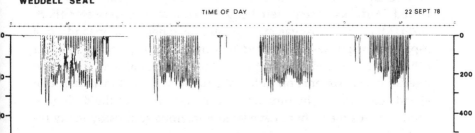

Figure 5.5. Computer plot of a free-ranging seal diving during the month of September when length of day is changing rapidly at McMurdo Sound. This animal was diving from dawn to dusk at this time of the month. The plot is of 4 days from a record of September 13 to 24. (From Kooyman et al. 1980.)

6

Physiology of diving

Festina lente–make haste slowly.
Augustus Caesar

When one learns of the extraordi-
nary capacities of marine mammals, some of the first questions are:
How do they hold their breaths for so long, and how do they tolerate
such great pressures? There is much active research on these questions, and although most of it has consisted of laboratory studies on
other species of marine mammals, the principles are believed to be
universal and apply to the Weddell seal as well. The contribution that
the Weddell seal is making to this body of knowledge is that it is the
only marine mammal on which studies under natural conditions are
presently feasible. Such studies are giving us more insight into the
relevance of some physiological responses observed in laboratory experiments and their significance in the normal activities of the animal.

The effects of submersion

The physiology of breathholding
has been a subject of experimentation since at least 1870, when Paul
Bert became curious about the heart slowing (bradycardia) that occurred in ducks when their heads were forced underwater (Bert,
1870). Much progress was made about the general understanding of
asphyxial responses in the 1930s, when Laurence Irving and Per
Scholander, first independently and then collaboratively, conducted
a series of experiments on different aquatic birds and mammals that
elucidated many of the fundamental characteristics of the diving response. Since then, there have been numerous studies by many investigators using different sophisticated techniques on various aspects of this response. There are several excellent reviews on this
subject (Andersen, 1966; Elsner et al., 1966; Irving, 1964; Scholander,
1963, 1964), and my objective is not to add another, but to give a brief

outline of the response in order to make the experiments described for Weddell seals more understandable.

There are several known ways by which breathholding ability is enhanced:

1. The oxygen stores of the body are increased. Oxygen may be stored as a gas in the lung, in physical solution in the blood and tissue fluids, or bound to hemoglobin molecules in the erythrocytes or to myoglobin in the muscle tissue. In seals, the major store is carried in the blood (Fig. 6.1). The carrying capacity of Weddell seal blood is 1.60 times that of human blood because of the greater concentration of hemoglobin. The blood volume per unit of body weight is 150 ml per kg, or about double that of humans. The net result is that blood O_2 stores per unit of body weight in seals are 3.0 to 3.5 that in man. Because blubber is rather inert and represents 30 percent of a seal's body weight, a more realistic comparison might be to lean body weight. In this case, the difference from man is 5.3 times. Furthermore, the myoglobin concentration in seal muscle is about ten times greater than in man, which represents another substantially larger O_2 store.

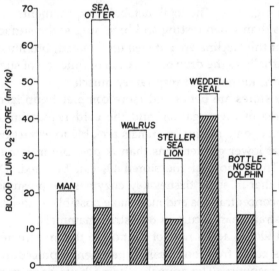

Figure 6.1. Blood and lung O_2 stores of the Weddell seal and other marine mammals. The hatched portion of the column represents the blood O_2 store, and the clear portion is the lung store. The blood store was calculated assuming that 33 percent of the blood volume was arterial and that it was 95 percent saturated with O_2. The remaining blood volume was venous, which held 5 volumes percent less O_2 than the arterial. The lung O_2 store was considered to be 15 percent of the total lung capacity. (From Kooyman, 1973.)

2. The oxygen stores of the body are rationed. The various organs of the body have different oxygen requirements and different abilities to function through anaerobic metabolic pathways. During a dive, blood flow changes dramatically in some organs and only slightly in others. The brain is a good example of an aerobic organ; it must keep functioning at full capacity during a dive, and there is little change in blood flow to this organ. In contrast, blood flow to the kidney nearly ceases during a dive and much of the kidney's filtering process stops until breathing is resumed. Blood flow to most of the muscles also declines, but in the swimming or struggling animal (because most studies are forced dives) the muscles continue to function. Their energy is believed to come via aerobic pathways initially, transitioning to anaerobic pathways later. These changes in blood flow can occur almost immediately upon submersion. The shifts in flow are reflected in a fall in heart rate as the restricted distribution of blood decreases the output requirements of the heart. The controls for these flow changes are orchestrated so exquisitely that little change in blood pressure occurs.

3. Energy requirements are less during a quiet (nonstruggling or swimming) dive. The total metabolic requirements of the animal are less than when resting and breathing at the surface. The reasons for this decline are not well understood, but must be at least due partially to the drop or cessation of function of such organs as the heart, kidney, and respiratory muscles.

4. Energy stores are better and more completely utilized. This is a new area of investigation, and the evidence accumulated indicates that some marine mammals are able to extract oxygen from blood at lower gas tensions than is possible in terrestrial mammals. Thus, more of the stored O_2 can be used. Others have shown that in some tissues key enzymes are presumed to be in higher concentrations and may make possible a more efficient use of substrate than occurs in terrestrial mammals.

I would like to add a note of caution to this minireview. Most of these facts come from one basic experimental procedure that must affect the response: The animals studied have been restrained and forcibly dived. The reasons for this approach are understandable. The animals are large and difficult to manipulate, the equipment necessary for measuring the physiological variables is bulky and requires that the animal stay in place, and those few variables that can be monitored by remote sensing equipment are limited in both the information they provide and the distance over which the signals can be received. Finally, if the studies are to be done in a natural environment there is nothing to stop the animal from leaving unless it is well-trained or forced to return for some reason. Enter the Weddell seal!

The Weddell seal is an ideal model (perhaps *the* model) on which to conduct unrestrained experiments for all the reasons discussed in Chapter 2. Added to the incentive of working with an unlimited number of untrained animals in *its* environment is the seal's great diving ability. We can begin to estimate the seal's diving ability by making some interesting calculations in which we assume that the body weight of the seal is 450 kg. (If you, the reader, are not inclined to play with numbers, skip seven paragraphs to my conclusion.) The lung volume during a dive is equivalent to 0.027 liters per kg of body weight, and 15 percent of this is O_2 or 1.9 liters of O_2 is in the lungs at the beginning of the dive. The blood volume is 67 liters, of which I assume that 23 liters are arterial and 95 percent saturated. The O_2 carrying capacity is 35 volumes percent, or 35 ml of O_2 per 100 ml of blood. It is assumed that O_2 can be extracted to 20 percent saturation, or 20 mm Hg O_2 tension. This yields 5.9 liters of O_2 available in the arterial supply. Assuming an arterial-venous difference of 5 volumes percent, and that extraction of venous O_2 is to 7 percent saturation (10 mm Hg), this means 11.3 liters of O_2 are available for consumption in this blood store. Thus, a total of 19.1 liters of circulatory O_2 are available during the dive. If we assume that muscle myoglobin saturates at 8 volumes percent and 30 percent of the body mass is muscle, then there are 11 liters of O_2 stored in muscle at the beginning of a dive. The total store is 30 liters.

If the oxygen consumption rate ($\dot{V}_{O_2}$) is 250 ml O_2 per kilogram per hour (an average value obtained from the respiration studies discussed in Chapter 7), this means a total $\dot{V}_{O_2}$ for a 450 kg animal is 1.9 liters per min. Then the 30 liters of O_2 available in the body will be consumed in about 16 min. We know that the O_2 store is not consumed in so short a time because they can breathhold for at least 73 min. These simple calculations illustrate how important the diving response is for extending the breathhold.

Now let us consider a well-orchestrated, full-blown diving response. In this case, O_2 bound to myoglobin should not be considered because it is sequestered in the muscles, does not circulate, and is used up rapidly. Therefore, a total of 19.1 liters of circulatory O_2 are available. With these calculations, we can arrive at a hypothetical maximum breathhold limit that is interesting to compare with the longest known dive duration. During an extended dive, the seal becomes a heart-brain-lung aerobe, in which circulating blood is limited to these organs. All others derive their energy from aerobic pathways until the O_2 stored in the organ is exhausted, and then they function anaerobically for the remainder of the dive. (This is an oversimplification, but it will yield an estimate of the maximum possible length of a breathhold and establish a basis for discussion.)

The $\dot{V}_{O_2}$ of the heart and brain has not been measured. The heart weight is 0.5 percent of the body weight, which is similar to that of other mammals. If we assume that it has a similar consumption rate per weight to the exponent of 0.75 (weight $^{0.75}$), then its $\dot{V}_{O_2}$ is about 5 percent of the total resting $\dot{V}_{O_2}$, or 95 ml of O_2 per min. The heart rate decreases 40 percent during the dive, which may lower its O_2 needs, but for simplicity this unknown is ignored. The brain weight is 0.13 percent of body weight and consumes 13.6 ml O_2 per min., or 0.7 percent of resting $\dot{V}_{O_2}$. As you might guess, animals with large brains in proportion to body size would make poor long-duration divers.

The Weddell seal, with a brain one-third the size of man's, probably consumes one-sixth the amount of O_2 according to $\dot{V}_{O_2}$ per weight to the exponent of 0.75 (weight $^{0.75}$).

The lung $\dot{V}_{O_2}$ is not known, but is presumed to be small under breathhold conditions. Therefore, the $\dot{V}_{O_2}$ of a heart-brain-lung system is estimated to be 6 percent of the resting $\dot{V}_{O_2}$, or 115 ml of O_2 per min. Thus, 19.1 liters of O_2 divided by 115 ml of O_2 per min. equals 166 min.

The Weddell seal could breathhold for 166 min. if it were a perfect diving machine. Of course, the seal is not perfect, and these calculations fail to take into account all regions that require some of the O_2 store during the dive; for example, the spinal cord. If the percent of resting $\dot{V}_{O_2}$ is plotted against the calculated dive duration, some interesting relationships are illustrated (Fig. 6.2). A distinct flexure occurs in the curve when the diving $\dot{V}_{O_2}$ becomes less than 20 percent. At this flexion, the relationship changes rapidly. If an animal can reduce its diving $\dot{V}_{O_2}$ below this level, it gains a lot in breathhold ability for small reductions in $\dot{V}_{O_2}$ if it can tolerate the change in tissue metabolite concentration and pH.

At the risk of tedium, I end with one more calculation. The seal that made the 73 min. dive weighed 450 kg, therefore, 19.1 liters divided by 73 min. = 262 ml of O_2 per min., or 14 percent of the assumed resting $\dot{V}_{O_2}$ of 1.9 liters per min. The values of 6 percent of resting $\dot{V}_{O_2}$ must be low, and I suspect that 10 to 15 percent is more realistic. This suggests that the Weddell seal's breathhold limit may be as much as 95 min. Importantly, the seal upon departing and returning from these long dives seems to be swimming at a very leisurely rate relative to deep dives of short duration. This may be an efficient energy conservation strategy: Water is kept flowing over the body at a low enough rate so that turbulence and drag are minimal. Perhaps this is a biological example of Augustus Caesar's admonition: "Make haste slowly."

Every time a seal dives, is a full-blown diving response elicited? If one considered only the laboratory studies of restrained dives, the answer would be yes. To me, this seemed unlikely and prompted an effort to get some evidence from Weddell seals.

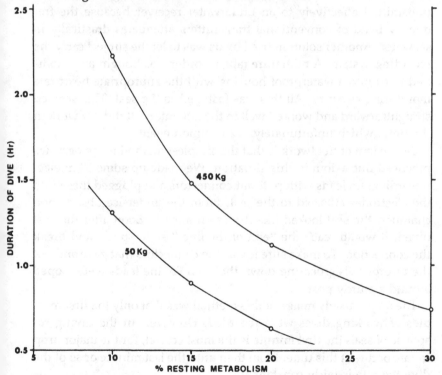

Figure 6.2. Calculated breathholding limit of a Weddell seal based on the O_2 stores of the body and the O_2 consumption rate during the dive. The two curves are for 50 and 450 kg seals. (From Kooyman, 1975.)

The heart rate is one of the more easily measured physiological variables; it is quite informative about the onset and profundity of the diving response. When seals are sleeping out of water, the beat sends a visible tremor through the blubber of the chest, and the rate may be determined easily. Between dives, the beat can be heard with a waterproofed stethoscope placed against the chest. Some seals will tolerate this procedure if done *carefully*, which is advisable because you are nose to nose at this time. More difficult, but still practical, surface electrodes can be attached to shaved areas of the skin and the electropotential differences generated by the heart muscle contractions can be recorded.

By measuring the seals' heart rates while they were voluntarily doing their different kinds of dives, comparative information could be obtained on the variability of the response. To do even this seemingly simple measurement in the uncontrolled conditions of the open sea raises major technical and cost difficulties. The signal cannot be transmitted effectively to an underwater receiver because the frequency band of conventional transmitters attenuates drastically in seawater. Another solution tried by us was to let the subject carry the recording system. A miniature tape recorder was bought and modified to fit into a waterproof housing with the appropriate heart rate monitoring circuitry. All this was fastened to the seal. This seemed straightforward and worked well in the laboratory. It did not work in the field, which unfortunately is a common event.

One axiom of fieldwork is that the simplest method is the best; we practiced this axiom in this situation. We made up some 80 m electrocardiogram leads with pull-out connectors and plugged these into the electrodes attached to the seal. From the under-ice observation chamber, the seal looked like it was on a leash. Soon after the seal dived, it would reach the "end of the line," so to speak, and break the connection. To make sure it was the connectors that gave and not the recorder disappearing down the ice hole, the leads were looped around a sturdy post.

The major disadvantage of this method was that only the first minutes of working dives were recorded. However, in the diving response of seals the first minute is the most crucial, for the major drop in rate occurs at this time. From then until the last minute or so of the dive the rate is fairly constant.

When an adult is sleeping on the ice surface, or in seawater at −1.9°C, the rate is about 50 to 60 beats per min. (BPM). This is a rough average because when seals are resting their breathing is cyclical; several breaths and then an apnea sometimes of several minutes. The heart rate varies with these cycles, slow during apnea and fast at eupnea. Averages during these periods are: apnea, 34 BPM; eupnea, 64 BPM. The rate between dives averages 85 BPM.

During dives, the rate varied according to the type of dive. During short and shallow but active dives, and dives where the seal was not active but resting just below the ice, the rate was 30 BPM. (We obtained similar rates from two eight month old Weddell seals at the Scripps Laboratories that were trained to remain submerged while chasing fish thrown into various parts of the pool.)

There was a statistically significant negative correlation with dive duration (Fig. 6.3). The longer the dive, the slower the heart rate, until at about 25 min. a minimum of 15 BPM was reached. Rates did not get any lower with longer dives. These results are interesting because they demonstrate that the diving response is not an on–off reflex, but is a complex response that is influenced and responds to the seal's anticipated plans. In other words, what is usually considered an organ under autonomic control is somehow being influenced by voluntary activity. Also, this experiment demonstrates that those studies using restraint in their procedures probably have a bias.

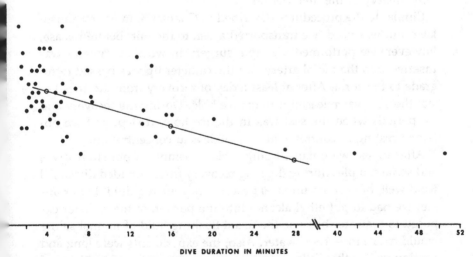

DIVE DURATION IN MINUTES

Figure 6.3. Relationship of heart rate to dive duration in adult Weddell seals. Regression line calculated from Bartlett's regression analysis, probability $(P) = 0.98$ that slope $(b) < 0$. (From Kooyman and Campbell, 1972.)

They are measuring the most profound response as the seal prepares for the worst: a submersion of unknown duration.

In nature, such profound responses are probably rare. Seals seldom make submersions that challenge their breathhold ability, except for those like the Weddell seal that must travel long distances under ice. Perhaps this is not surprising if we view it another way: As adults, most of us rarely if ever run as fast as we can.

Anticipated termination of the dive has similar but opposite effects on the heart rate. Those dives that we were able to record in their entirety showed an acceleration of heart rate shortly before the seal surfaced. Results are similar, in this case, during forced dives. When both Weddell seals and elephant seals were forced to dive and compressed in a chamber to simulate a deep dive, the heart rate almost always increased as the pressure was reduced. In one instance, the pressure was reapplied and the heart rate declined again. The heart rate variations provide tantalizing evidence of what may be happening in a cardiovascular sense during the dive. To learn more about the responses of seals to diving, a group of us measured the lactic acid concentration in arterial blood of adult seals (Kooyman et al., 1980). We did this while they were resting and after dives of durations varying from 10 to 61 min. The procedures were not easy, but well worth the effort because lactate yields an informative metabolic history of the dive period.

Similar to the procedures described in Chapter 5, in which an isolated hut was used, we transported a seal to the hut. Before release, however, we performed a simple surgery in which a catheter was inserted into the radial artery and the catheter tip was passed retrograde to the aorta. After at least a day of recovery from the anesthesia, the seal was released into the ice hole. During the recovery and for periods when the seal was in the ice hole, it slept and we obtained resting or control values of lactic acid concentration.

After dives, we obtained single blood samples from short dives and serial samples during the long recovery from extended dives. All went well, but certain unusual procedures were required. For example, we had to put ethyl alcohol into the portion of the catheter exposed to seawater because the usual flushing fluid of normal saline would freeze in − 1.9°C water. Also, the experiments were long and continuous, so that little sleep was obtained over the 48 to 72 hr. of the experiments. There was even some grumbling about being tyrannized by a seal from Eric Wahrenbrock, the anesthesiologist; Ev Sinnett, Harvard postdoctoral fellow; and Randall Davis and Mike Castellini, graduate students.

The results were exciting (Fig. 6.4). The plot shows that no significant increase occurs in lactate until the dive exceeds 20 min. For the first time, to my knowledge, the aerobic limit of an actively diving vertebrate is defined. Based on my earlier calculations, in which the O_2 stores would last 16 min. in a resting animal, this shows that in the actively diving seal they last longer! Obviously, there is some conservation of oxygen by shunting flow away from some organs. This is corroborated by the mild decline in heart rate for the shorter dives. Those organs of reduced flow simply operate at a lower metabolic rate and thus with lower oxygen demand.

Another significant point is that the aerobic dives result in more

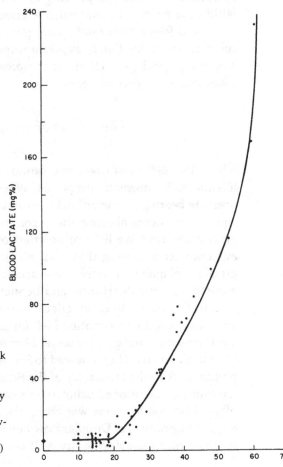

Figure 6.4. A summary of the peak concentration of lactic acid occurring in the arterial blood of three adult Weddell seals after voluntary dives of different durations. The diamond on the ordinate is the average resting lactate of the three seals. (From Kooyman et al, 1980.)

efficient use of time. From our study, it was clear that the longer dives require a protracted recovery; thus a 45 min. dive results in a surface period of about 60 min. Out of a total of 105 min., 43 percent is spent diving. If the seal makes five 15 min. dives, of which each requires 4 min. at the surface recovering, then over a 95 min. period 79 percent of the time is spent diving.

The evidence from free-ranging seals (Chapter 8) is that they prefer aerobic dives. Out of about 4,600 dives measured, only 3 percent of the dives were in excess of 25 min., the dive duration beyond which the lactate load from anaerobic glycolysis increases rapidly (Fig. 6.4).

Based upon evidence such as the heart rate and the lactic acid levels, I conclude that the dive response should not be called a "reflex" as is sometimes done. There is probably a wide range of gradations in the response, depending upon a variety of stimuli upon and within the animal. In contrast, the effects of compression are more direct and fewer compromises are possible. Indeed, the mention of compression immediately evokes a response in me to begin discussing the physiological effects of compression – a subject that I have investigated for several years.

The effects of compression

After measuring the depth to which Weddell seals dive, one cannot help but wonder how they tolerate such enormous pressures. When a seal reaches 600 m, the pressure bearing on the animal is 60 atmospheres, or about 64 kg per square cm. Remembering the incompressibility of liquids, these great pressures have little or no effect on most organs, but there is evidence accumulating that from about 50 to 60 atmospheres and greater, anomalies in nerve conduction begin to be manifested in terrestrial mammals (Hunter and Bennett, 1974).

With this desire to study effects of compression in mind, shortly after finishing my dissertation I left for England to study under Richard Harrison. During a year there, I became versed in the anatomy of marine mammals. Then I moved to Scripps Institution of Oceanography to work in the laboratory of Per Scholander and Robert Elsner. I have already mentioned Scholander's contributions to diving physiology. Elsner at this time was one of the few physiologists using the elegant techniques of Doppler flowmeters to quantify the blood flow to different organs during dives. It was here with their advice and help that I began to build the equipment necessary for pressure studies. There had been few such studies done and it was clear why. They require specialty items that are heavy, bulky, and expensive. Within a year, the research support shops had built a hydraulic compression chamber for me capable of 50 atmospheres pressure. But more about the equipment, procedures, and experiments after discussing three physical laws that are of paramount importance to understanding the problem of breathhold diving to depth.

In 1660 Robert Boyle noted that the volume of an ideal gas was inversely related to pressure, $PV = C$ (pressure times volume = constant), when the temperature remains constant. If the total pressure increases, then the partial pressure of each gas within a mixture must increase also, and, in 1801, John Dalton set forth this as occurring in equal proportions. Thus, $P_T = P_1 + P_2 + P_3 + \ldots$. William Henry in 1803 proposed that the solubility of a gas in solution is nearly proportional to its partial pressure in the gas phase at a given temperature. Therefore, $V_d = \alpha P V_L$, where V_d is the volume of the gas dissolved, V_L the volume of the liquid, P the pressure of the gas over the liquid, and α the solubility coefficient of the gas at a specific temperature.

The implication of Boyle's law is that the deeper the seal descends, the smaller the lungs, trachea, middle ear cavity, and any other gas spaces must become. I have no doubt that this must occur and an equilibrium between external and internal pressure must be maintained. If this were not so, membranes and blood vessels lining the cavities would rupture and obliterate the space with blood and cell fluids because the body fluids will always nearly match the ambient hydrostatic pressure. When an inequality occurs, much pain would probably be experienced. A good example of a disequilibrium in a gas space limiting the depth of dive is man. When Albert Croft made his record free dive to over 60 m, pain in the middle ear prevented him from going deeper (Schaefer *et al.*, 1968). The middle ear space of man is surrounded by bone and is incapable of shrinking much. For equilibrium to persist, air must be continually forced into the space. Any scuba diver is well aware of this and the effort he or she must continue to make to maintain equilibrium. Because Croft was free diving, 60 m must have been the point where there was no longer enough air in the shrinking lungs to force some into the middle ear.

Anatomical studies have elucidated how the seal manages this problem (Odend'hal and Poulter, 1966; Graham, 1967). The middle ear is lined with a system of venous sinuses. When the seal descends, the absolute venous blood pressure begins to increase and this additional pressure expands the sinuses. This causes the middle ear membranes to bulge into the middle ear space, reduce its volume, and compress the gas in the space. The result is a close match to the ambient pressure. This system probably requires no additional gas to be supplied from the lungs through the eustachian tube (air duct running from the middle ear to the back of the mouth), and if the tube is blocked for various reasons the seal can still dive deeply. What is not understood is the effect great reductions in the middle ear volume have on transmission of sound from the tympanic membrane through the ossicles and within the cochlea.

That a seal can continue to dive deeply even if the eustachian tube is blocked must catch the attention of any scuba diver. Without patent air ducts in the head, usually caused by a simple head cold, he or she is helpless because not only is it impossible to maintain pressure equilibrium in the middle ears, but the several air sinuses of the head (frontal, nasal, ethmoid) cannot be equilibrated either. The function of bony head sinuses is not clear, but whatever the need they are a liability to diving. Apparently, their function is not essential to mammals in the marine world because seals and sea lions do not have them. The result – elimination of an equilibration requirement.

The respiratory system is the largest gas space of all, and obviously it has great volume-changing properties – otherwise we could not breathe. In terrestrial mammals, there is a limit to the amount of gas that can be forcibly removed from the lung. This is called the residual volume, and in man this makes up 25 percent of the total lung volume. At this volume, the ribs are incapable of contracting further and the thoracic volume is fixed. There is another way to continue to reduce the lung volume. Thoracic and abdominal veins can swell with blood and further reduce the volume within the rigid chest. The latter may do so by pressing against the relaxed diaphragm, or by swelling within the thorax. This must have happened in Crofts' case because he dived beyond the depth where it was calculated that his diving lung volume would have been compressed to residual volume. If some such kind of compensation had not occurred, Croft should have suffered from edema in the lungs, as well as rupturing of pulmonary capillaries and bleeding into the lungs. There was no report of these adverse effects.

Seals have no such limitations. The thorax is more compressible because of more flexible, indeed rather rubbery ribs. The rubbery character of the ribs is due to a high proportion of cartilage invested in these bones. This resilience, as well as some peculiar properties of the bronchioles, allows nearly all the gas in the lungs to be exhaled. This absence of a residual volume was first noted by Scholander (1940). He knew that when the thoracic wall of any recently dead terrestrial mammal is punctured there is a sigh as the lungs deflate more completely when freed from the constraints of the rib cage. The residual volume is due to an equilibrium between the elastic lungs and the rigid chest wall. The lungs are prevented from recoiling further because of the slight negative pressure within the chest cavity and because surface tension of the liquid interface between the lungs and chest wall causes adherence between the two. When the chest is punctured (pneumothorax), this bond is destroyed, the lungs recoil completely, and more air is exhausted. However, collapse of the lung freed from the constraints of the chest is limited, and even then there remains a small volume of gas in the lungs.

Scholander noted that when a pneumothorax was created in a porpoise cadaver often no gas was expired. Further dissection showed that the lungs were completely collapsed. This novel property of the lung and chest wall has now been observed to be characteristic of several species of marine mammals in as diverse groups as large whales and seals.

In one characteristic, the Weddell seal and two other species of antarctic seals, the leopard and Ross seals, seem to have gone a step further than most in collapsibility of the respiratory system. In these three species, even the trachea is exceptionally pliant, so much so that when relaxed the lumen is nearly obliterated. This could permit these animals to dive to depths where compression is so great that even the major airways collapse.

To gain some direct evidence of what portions of the respiratory system are compressing the most, Douglas "Ted" Hammond, at the time a veterinarian in our laboratory, and I decided to make some radiograms of a young Weddell seal at both the surface, similar to what had been obtained for harbor seals, *Phoca vitulina* (Fig. 6.5), and when it was in a simulated dive to 300 m. Although the experiment was theoretically simple, the methods necessary to accomplish the task were a bit complicated and novel. The X rays would first have to penetrate the steel compression chamber with its 1 cm thick steel walls and then 46 cm of water and seal. The only unit

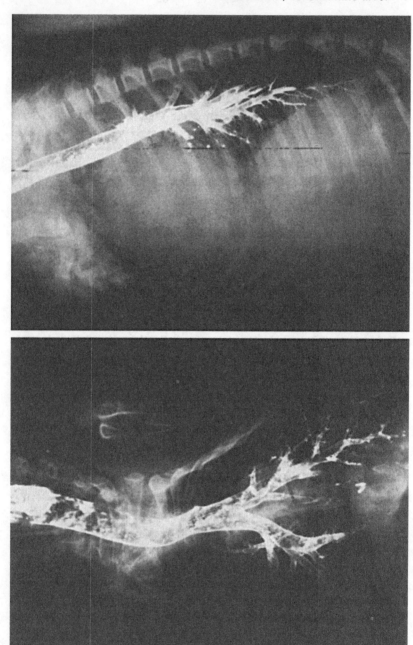

Figure 6.5. Radiogram (*above*, lateral; *below*, dorsal) of the trachea and bronchioles of a harbor seal, *Phoca vitulina*. Contrast of the airways has been enhanced by previously insufflating them with tantalum. (Radiogram by D. D. Hammond.)

that we knew to be powerful enough to even come close to doing the task was one in the radiology department at UCLA. This was fortunate because at the time Ted was taking a specialty course in radiography. According to Ted's calculations, we would have to reduce the amount of material the X rays had to penetrate. A 5 cm diameter by 0.5 cm thick aluminum window was put in so that lateral and dorsal-ventral radiograms could be taken. Tests were made to see if the film could tolerate the pressures. Serendipitously, we found it could not only withstand the pressure without damaging the emulsion, but it became more sensitive. The X ray film and intensifying screen were placed in a watertight plastic jacket and wedged against the seal, so a method had been worked out. Now all that was required was to get a 900 kg chamber, three 200 liter drums for water reservoirs, and all sorts of other paraphernalia, including a 90 kg seal, from Scripps to UCLA. This all had to be done on a weekend when the radiology unit was not in use for more mundane matters such as patient limb fractures.

On a Friday afternoon, my technician left Scripps with a well-laden university truck, loaded and tied down by amateurs. Ted and I were to follow, and we made a pact that if a compression chamber were seen scattered about on the San Diego Freeway going toward Los Angeles we would drive on as if we knew nothing about it.

Our arrival at UCLA was like a commando raid. Everything had to be planned precisely, including getting a truck whose bed matched the height of the loading dock, so that the chamber could be removed without the help of a crane. Late that evening, as we wheeled the chamber with its ominous, low rumble down the long hospital corridor, we passed two interns chatting in the hallway. To my surprise they took no notice of us, the chamber, or the seal. I still have not decided whether this was because interns are so intensely involved in medical matters or are frequently half asleep.

The system worked, and for all that effort we obtained several X rays of the trachea, bronchioles, and lungs, one of which can be seen in Figure 6.6. The evidence obtained from this experiment led us to the conclusion that there is a graded collapse of the respiratory system as it is subjected to increasing ambient pressures. Even at 30 atmospheres, we could see no significant change in the size of the bronchioles; but the tracheal diameter had changed much. Significantly, the tracheal lumen was still patent and could continue to function as a sound-resonating structure. The tracheal volume de-

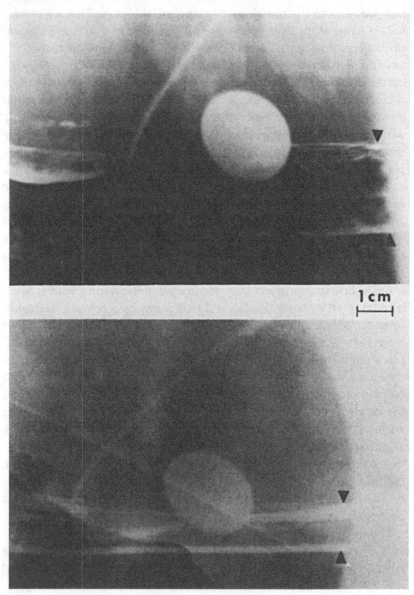

Figure 6.6. Radiograms of the lateral portion of the trachea of a Weddell seal just posterior to the glottis. The upper photo was taken when the seal was submerged at 1 atmosphere absolute (ATA). The lower picture was taken when the seal was submerged and compressed to 31.6 ATA. Anterior is to the left; arrowheads define the boundaries of the trachea; the circular object is a nickel that was used in an electrode contact for monitoring the electrocardiogram. (From Kooyman et al., 1970.)

creased to one-fourth of its original volume during the 30 atmosphere dives, instead of to one-thirtieth as it theoretically should have if it had been an isolated structure. Therefore, gas must have come from other sources. The only other sources of sufficient volume are the alveoli, which in their turn must be more pliant and collapsing at a more rapid rate than if they had been isolated units of the lung.

At first, this might seem to be a maladaptive property – that the gas-exchanging regions of the lungs would lose gas, and therefore oxygen, to the nonexchanging upper airways. However, it should be kept in mind that with a four times greater concentration of N_2 than O_2, the lung is a better N_2 than O_2 reservoir. According to Dalton's and Henry's laws, as the seal descends, the total pressure of the lung gases rise, and both the partial pressure of each gas and the solubilities of the gases increase almost linearly. As deep as the Weddell seals dive, there is a potential for large amounts of N_2 being dissolved in the blood and tissue. There, exposures could result in N_2 narcosis and the bends. Humans breathing compressed air experience the symptoms of narcosis rapidly after exceeding a depth of about 40 m. The symptoms are impairment of judgment, of neuromuscular coordination, and of vision – not desirable qualities for an animal seeking fish or breathing holes.

The occurrence of the bends is dependent on the depth of the dive, the exposure time, and the rate of ascent. (The term *bends* came from the symptoms of the diver doubling up due to pain in various joints.) For human divers, tables have been developed that tell a diver how deep he can go and how long he can stay without suffering the bends. The deeper the diver goes, the higher the partial pressure of N_2, the greater the N_2's solubility in blood and tissue fluids, and the faster the N_2 is absorbed; therefore, the less time the diver can stay at depth. If the diver exceeds the prescribed limits, he may suffer gas bubble formation in the tissue and circulatory system, or the bends. Other results of the bends can be permanent paralysis or death due to a variety of causes.

For some time, it was thought that the bends were only a liability in the case of divers breathing compressed air, because when breathhold diving the N_2 supply is limited and there is not a sufficient amount to raise N_2 tension to a dangerous level. However, a Norwegian medical officer and diver has described a special case of free diving where the symptoms are similar to the bends (Paulev, 1965). These symptoms occur after the diver makes repetitive dives to 20 m over a period of several hours with only a short surface interval in between.

Weddell seals engage in much more demanding diving schedules (Fig. 5.5). But because of the depth and duration of some of their dives, the bends could be contracted even on a single dive if there were not some protective mechanisms. This conclusion is based upon calculations summarized in Table 6.1.

We know from behavioral studies that the seals dive long enough and deep enough to absorb all the N_2 in the lungs. We also know that they ascend too rapidly for dissolved gases to come out of solution safely in the lung. Furthermore, the circulatory shifts that occur when a seal dives, especially for a long time, distribute the blood flow in the worst possible way for an animal whose lung N_2 tensions are soaring. Flow is restricted to the heart and nervous system; thus, any N_2 taken up by the blood is limited to only a few organs of the body, and these organs are some of the most susceptible.

As you can see from Table 6.1, the N_2 tensions could rise as high as 7 atmospheres, depending upon blood flow distribution. The tensions could certainly rise high enough to cause narcosis on the way down and gas bubble formation on the way up. To determine the actual N_2 tensions to which the tissues are exposed, arterial and venous blood samples were obtained from several elephant seals and one Weddell seal when they were compressed to various depth

Table 6.1. *The N_2 tension that would occur in an adult Weddell seal if all the lung N_2 were absorbed during a deep dive, and the dependence of the tension on how it is distributed within the body.*

N_2 distribution	N_2 tension
Blood	10.5
Blood + 20% extravascular H_2O	7.4
Total body H_2O	3.7
Total body H_2O + fat	1.5

Assumptions: Body weight = 425 kg; diving lung N_2 volume = 7.7 liters, standard temperature and pressure, dry; blood volume = 63 liters; body fat = 128 kg; total body water = 208 liters; N_2 solubility in blood = 14 ml/liter/atm of N_2; N_2 solubility in fat = 70 ml/liter/atm of N_2; alveolar N_2 fraction = 0.9; diving heart rate = 10/min.; and diving cardiac output = 10 liters/min. *Source:* Adapted from Kooyman, 1972.

equivalents. Elephant seals were selected for this study rather than Weddell seals because it was simpler to take the chamber to Guadalupe Island, Mexico, aboard the research vessel *Alpha Helix*, than to the Antarctic; or than to bring more seals from the Antarctic. Morphologically and behaviorally, the seals are similar enough that there should be no significant difference in results between the two.

No matter what pressure was applied above 4 atmospheres absolute (ATA), the arterial and N_2 tensions remained at the same low value of 2 to 3 ATA (Fig. 6.7). There is probably too small a difference from the normal surface tension to cause bubble formation, no matter how fast the seal ascends. These arterial and venous N_2 tensions are not high enough to be in accordance with the amount of N_2 available for absorption in the lungs. Also, the close match between arterial and venous N_2 tensions soon after compression is evidence that no further N_2 uptake from the lungs was in progress; otherwise, arterial N_2 tension would continue to be higher than venous N_2 tension.

A significant advantage for a graded collapse in the lung and its airways becomes clear. Collapse of the alveoli before much of the gas within them diffuses into the blood prevents high N_2 tensions in the blood and tissue. The seal can dive as deeply and as long as it desires and no further gas exchange will occur because the lungs are collapsed. A model of how this lung collapse progresses during a dive in seals is diagramed in Figure 6.8.

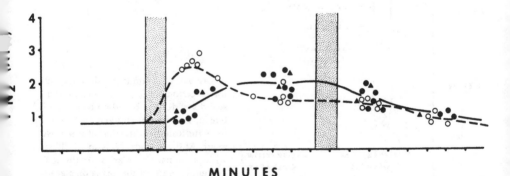

MINUTES

Figure 6.7. Arterial and venous N_2 tensions of elephant seals and one Weddell seal (triangles) after submersions and compressions ranging from 4 to 14.6 ATA. Arrows indicate when the dives began and ended. Hatched columns are the compression and decompression intervals. Open circles are arterial values. Closed circles and triangles are central venous samples. All values are the averages of two or three samplings. Arterial and venous samples were obtained during separate experiments from different animals. (From Kooyman et al., 1972.)

It is interesting that seals consistently exhale to a low lung volume before diving. This insures that alveolar collapse will occur at a shallow depth. The seal does not know this and must exhale for other reasons, but it has this fringe benefit.

This experimental evidence supports a hypothesis proposed thirty years earlier by Scholander, in which he based his argument on morphological studies (Scholander, 1940). Two important observations caught his attention: (1) The lungs of marine mammals empty more completely than those of terrestrial mammals; this was discussed earlier. (2) This feature was possible because of the unusual structure of marine mammal airways. The bronchioles are reinforced with cartilage and muscle more extensively than those of terrestrial mammals. This provided them with a more rigid airway system that would not occlude at low lung volumes. Thus, if the lungs were compressed to a small volume during a deep dive, the airway diameters would remain constant while gas was being forced out of the shrinking alveoli and into non-gas-exchanging upper airways.

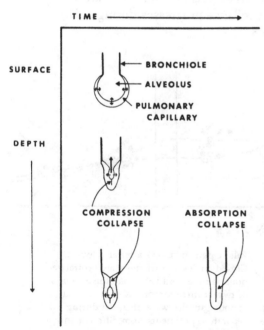

Figure 6.8. Diagram of an alveolus and the direction of N_2 gas movement as a seal dives deeply; based on measured N_2 tensions in arterial and venous blood. Arrows indicate the direction of gas movement. At the surface, N_2 is in equilibrium with pulmonary capillaries. As the seal descends, much of the gas is displaced into the bronchioles and bronchi. At some depth, probably less than 50 m, a small amount of gas is trapped in the alveoli due to closure of the airway. This gas will be absorbed if the seal remains at depth long enough. (From Kooyman et al., 1972.)

Human medicine

The unusual physiology of seals has attracted the interest of investigators who are usually involved in problems of human medicine. It has been learned that the diving response of seals that is so obvious in this group also occurs in every other vertebrate, including man. However, the response is subtle enough in nondivers that it is easily overlooked. Also, it is so short term (most humans cannot safely hold their breath beyond 2 min.), that the response cannot be studied in the detail possible in seals. Nevertheless, the response is universal; even occurring in the breeding grunion and the flying fish, which leave the water for brief periods during which they are essentially breathholding. Significantly, it occurs when a baby is making the transition from aquatic life in the womb to terrestrial life, where it must breathe air. At this critical stage, it is important to understand that the brief period of asphyxia is similar to the diving response. If for some reason the asphyxial period is protracted, such vital signs as a profound bradycardia are normal and essential to the breathholding newborn and in some instances should not be overridden with stimulants.

The major reduction in blood circulation to certain organs during a dive has come to the attention of those concerned with organ transplants. It is known that the isolated kidney of the seal can recover and function after much longer periods without perfusion than that of the dog (Halasz et al., 1974). Comparisons between these two species showed that after ischemic episodes as long as 60 min. the dog's renal blood flow and V_{O_2} were one-fifth as much as before, and urine production had ceased. These results indicate a general failure of the kidney. After similar treatment to the seal's kidney, these same variables returned to preischemic values. The factors that make seal kidneys and other organs resistant to anoxia are not known, and are of considerable interest not only to the organ transplant specialist, but also to the cell physiologist and biochemist.

Thousands of newborn infants die annually of respiratory distress syndrome, a condition in which lung inflation is difficult and full inflation is not achieved; after each breath, the lungs deflate to nearly zero volume and the cycle begins again. Even in adults, lung collapse is a serious matter and reinflation requires substantial time and effort. In contrast, as a Weddell seal dives to depth many times a day its lungs must be near total collapse during many of these plunges.

74   *Weddell seal, consummate diver*

Yet recovery seems to be complete after each dive, and there are no aftereffects. Even during the abnormal conditions of anesthesia, when the chest musculature of the seal relaxes and lung collapse may occur, the seal has no trouble reinflating its lungs. This ability not to be discomfited by lung collapse is a remarkable mammalian feature that is not understood and continues to occupy much of my investigative time.

Suggested reading

Andersen, H. T. 1966. Physiological adaptations in diving vertebrates. *Physiological Review* 46:212–43.

Kooyman, G. L. 1972. Deep diving behaviour and effects of pressure in reptiles, birds and mammals. In *Symposia of the Society for Experimental Biology: The Effects of Pressure on Organisms*, vol. 26, pp. 295–311, ed. M. A. Sleigh and A. G. MacDonald. Cambridge: Cambridge University Press.

– 1973. Respiratory adaptations in marine mammals. *American Zoologist* 13:457–68.

Kooyman, G. L., and H. T. Andersen. 1969. Deep diving. In *Biology of Marine Mammals*, pp. 65–94, ed. H. T. Andersen. New York: Academic Press.

Scholander, P. F. 1963. The master switch of life. *Scientific American* 209:92–106.

7

Food habits and energetics

Feed me with food convenient for me.
Proverbs 30:8

The standard method to determine the food habits of many animals, especially those that cannot be watched while they are feeding, is to collect their stomachs and analyze the contents. Because marine mammals digest their food rapidly, the contents of their stomachs are almost always fragmented and well along in the digestive process. Consequently, the work of identification is tedious, requires much taxonomic skill, and few people are qualified for it. Because the procedure requires killing the animal, most biologists are not inclined to do this for the sake of stomach analysis only. As a result, only a few studies have been done and these were ancillary to other projects in which it was necessary to sacrifice the seal.

The most detailed study that I am aware of was done by my friend John Dearborn (Dearborn, 1965), who, as I mentioned earlier, showed me my first Weddell seal. According to John, fish occurred in about 97 percent of the stomachs of seals examined at McMurdo Sound. The species ranged from ice fish to antarctic cod, but most were antarctic smelt or belonged to a group of the genus *Trematomus*, for which there is no common name. Weddell seals in the Antarctic Peninsula area tend to feed more frequently on cephalopods (Bertram, 1940).

All these collections have been over short periods of time and there are no doubt seasonal and locality variations that dictate what the seals eat. For example, offshore in the midwater depths of McMurdo Sound they eat antarctic cod and smelt. We know the occurrence of the cod is seasonal; they appear in the Sound around October and disappear by late December. During this time, I was once impressed to see an adult seal catch and consume a fish of about 30 kg in 2 to 3 hrs; within less than 24 hr. it had caught and eaten part of another large fish (Fig. 7.1). The total amount of fish eaten in the 24 hr. was about 45 kg. The uneaten portion of the second fish was stored beneath the ice, next to the breathing hole. An even bigger fish story

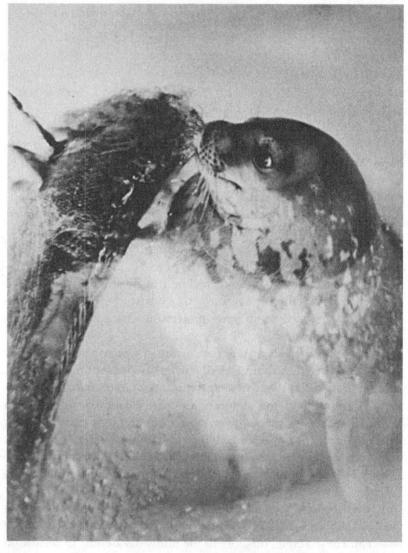

Figure 7.1. A Weddell seal that has just surfaced with an antarctic cod, *D. mawsoni*. (Courtesy P. Koerwitz.)

was reported by New Zealand scientists (Calhaelm and Christoffel, 1969). Over a 2 week period in December, 1966, they observed an intense feeding effort by one Weddell seal. The seal's catch per day was about 68 kg of *D. mawsoni*.

There have been a number of measurements of the metabolic rates of resting marine mammals, especially seals. Most have used an indirect procedure by determining the oxygen consumption ($\dot{V}_{O_2}$) of the animal. In all cases the resting $\dot{V}_{O_2}$ of marine mammals is 1.5 to 3.0 times more than the value for terrestrial mammals of similar size. (See Figure 4.6, which only presents data on seals; I will remain specific in this discussion.) Several species have been studied over the years, and the resting $\dot{V}_{O_2}$'s are summarized in Table 7.1 (these are the same data as in Fig. 4.6). Some data were obtained while the seals rested in a metabolic chamber, while in other experiments the seals were forcibly restrained during the measurements.

An exception to all methods was the procedure for adult Weddell seals. The data were collected while the animals were resting and

>le 7.1. *Resting metabolic rate of various species of seals.*[a]

cies	Average mass (kg)	Number studied	$\dot{V}_{O_2}$ (mlO$_2$·kg^{-1}·hr.$^{-1}$)	Predicted[a] $\dot{V}_{O_2}$ (mlO$_2$·kg^{-1}·hr.$^{-1}$)	Multiple of terrestrial mammal
·bor seal[b] *oca vitulina*	27.4	6	475	284	1.7
itulina	13.6	2	740	338	2.2
·nland seal[b] *agophilus*	11.4	6	1,048	354	3.0
oenlandicus)	31.6	3	444	274	1.6
·ddell seal ·eptonychotes ·ddelli)	65	5	408	228	1.8
eddelli[c]	425	5	309	143	2.2
·estrial mammal	70		225	225	

·e standard terrestrial mammal equation is: $\dot{V}_{O_2}/M_b = 650M_b^{0.25}$, where $\dot{V}_{O_2}/M_b =$ mlO$_2$· kg^{-1} · hr.$^{-1}$

M_b = body mass in kg.

·ues from restrained animals.

·m free-ranging animals resting in seawater at −1.86°C.

·rences: (1) Hart and Irving, 1959; (2) Miller, Rosenmann, and Morrison, 1976; (3) ·arova, 1964; (4) Elsner et al., 1977; (5) Kooyman et al., 1973.

sleeping in ice holes. A seal was required to breathe into a nose piece with two one-way valves. The overall arrangement is illustrated in Figure 7.2. The seals were willing to breathe into the valve after voluntary dives, as well as while they were resting or sleeping.

The experimental concept is a natural extension of the normal condition. The seals often breathe through small holes in the ice in which only their noses can be projected above the water level. It seemed a natural and simple matter to take advantage of these circumstances for respiration studies. Nevertheless, there were a few things to learn about seal preferences before the design was successful. Initially, a plywood sheet with a 6 to 8 cm diameter hole cut in the center was placed on the water. This covered the rest of the hole entirely so that the seal had to breathe in the precut orifice. The first seal tested had other ideas and immediately set to work enlarging the hole. When a 450 kg seal begins to ream plywood, splinters fly and the results are quickly devastating. By trial and error, we discovered

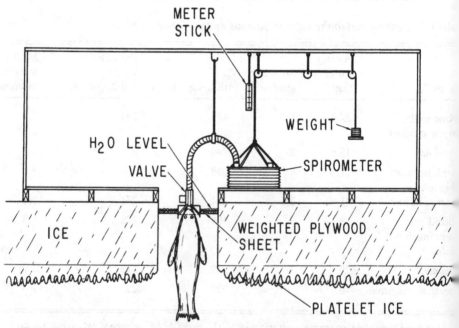

Figure 7.2. General arrangement of the respiratory apparatus inside the hut. The ice hole is completely covered with a weighted sheet of plywood, with the exception of a small hole in its center. This hole is covered with a lucite dome, upon which is mounted a valve with an inspir-atory and expiratory port. The seal's expirations are collected in the spirometer. The volume of the gas collected and its composition make possible the calculation of metabolic rate. (From Kooyman et al., 1971.)

Table 7.2. *Oxygen consumption rates of Weddell seals while resting in seawater at a temperature of* −1.86°C.[a]

Seal[b]	Sample size	$\dot{V}_{O_2}$[c] Mean	Range	Total collection time (hr.)
1	5	193	176–222	1.8
2	12	346	217–433	2.7
	7	240	225–260	2.0
3	5	335	180–424	1.2
4	4	257	175–315	1.2
5	8	345	273–480	2.2
	5	407	389–422	1.1
Mean		309		

[a]More than one entry represents analyses on different days. Collections were made after the seals had been sleeping soundly for a period of 1 to 2 hr.
[b]These numbers refer to the individual Weddell seals studied. All were adults and weighed 370 to 450 kg. Number 1 was a female.
[c]$\dot{V}_{O_2} = \text{mlO}_2 \cdot \text{kg}^{-1} \cdot \text{hr.}^{-1}$.
Source: Kooyman et al., 1973.

Figure 7.3. An adult Weddell seal breathing through the orifice in the top of the plastic dome. (Not shown are the valve and hoses that direct the expired gas to the spirometer.)

that if an animal could raise its entire head above water, and if the dome into which it rose was clear plastic so that it could see, then the seal was usually satisfied. The final design is illustrated in Figure 7.3.

When the seal rose into the small dome, it placed its nose so tightly into the orifice that an airtight union was formed. The rest periods sometimes lasted 8 or 9 hr., and gas samples were collected during the periods when the seal seemed to be sleeping most soundly. The average metabolic rate while resting in water at $-1.86°C$ was 309 ml O_2 per kilogram hour (Table 7.2), which is twice that of a terrestrial mammal of similar mass (Table 7.1).

In addition to rest periods, the seals would spend 6 to 8 hr. diving earnestly, meaning they were feeding or searching for other breathing holes. The rest of the time was spent in short and shallow dives near the hole. When it was believed that the seal had begun a regular diving routine for feeding, we would collect all exhaled gas samples as the seal came and went from the hut.

In an attempt to gain some information on the energetics of the seals while swimming, the overall metabolism over a few of the hours that they came and went from the breathing hole was determined. Presumably, at these times they were hunting for fish or for other breathing holes. The drawback to this experimental procedure is that it would not give us any information regarding cost of transport or cost of swimming at different speeds, because we were unable to determine the distance or speed that the seals were swimming. However, it seemed a start toward gaining some insight into the physiology of exercise and the energetics of marine mammals at sea.

The cost of diving results are perplexing (Table 7.3). The overall average $\dot{V}_{O_2}$, that is, the total oxygen consumed over a period of several hours of diving divided by the sum of the diving time plus the surface ventilation time, is slightly lower than the oxygen consumed while the animal is resting or sleeping at the surface! Such a result seems unlikely at first, and possible sources of error were considered. The analytical procedures could be faulty, but, if so, the error should occur in both the exercising and resting analysis. There seemed to be no fault in the analysis. Another possibility is that the seals were going to another breathing hole. This would cause a low value, as some of the oxygen consumption would be missed. However, for such an occurrence the seals' time away would be unusually long, yet the recoveries would be unusually mild. I am confident that no such incidences occurred, especially because in some cases a time-depth recorder was attached, and we could verify the dive durations. Finally, perhaps a more careful look at the original data is warranted (Tables 7.2 and 7.3). The results of resting animals suggest that at times they simply may not have been completely at rest. Because all data were averaged together, this raises the mean resting rate to a high value.

In contrast, once seals begin to dive, their routine may become less variable in terms of the energy output of the animal. This might be similar to a long-distance walk in which a constant energy output is established and maintained rather precisely, whereas a resting person's energy output may vary markedly depending upon his or her anxiety level, how often he or she gets up and moves about, and so on. Furthermore, these resting animals most likely had been feeding previously. Most digestion of food takes place during the rest period rather than during the dive, when blood flow to the viscera is limited (see Chapter 6). The specific dynamic action (SDA) of protein digestion can elevate metabolism from 10 to 30 percent. If conditions of the resting animals could have been more controlled (and less natural), it is likely that we would have noted lower metabolic rates. However, let me emphasize the most important finding of this study was that there is not much metabolism during the diving periods, and therefore the cost of diving and hunting is not great.

Table 7.3. *Oxygen consumption rates of Weddell seals during recoveries from a series of dives.*[a]

Seal[b]	$\dot{V}_{O_2}$[c]	Number of dives	Total surface time (hr.)	Total dive time (hr.)	Total time (hr.)
2	274	22	1.6	4.8	6.4
3	292	22	1.7	3.6	5.3
4	211	6	1.2	2.3	3.5
5	236	23	1.6	3.5	5.1
Mean	254				

[a]The rates were calculated by including both the surface time and the dive time.
[b]These numbers refer to the same individuals as in Table 7.2.
[c]$\dot{V}_{O_2} = mlO_2 \cdot kg^{-1} \cdot hr.^{-1}$.
Source: Kooyman et al., 1973.

This tends to agree with what has been shown in a variety of studies by a number of investigators. Swimming is the most efficient means of moving about, followed by flying and then running (Fig. 7.4). This may seem odd because water is such a dense medium, but little or no support of the body mass is required and all muscular effort can be devoted to propelling the animal forward. To reduce this effort, considerable and similar streamlining has occurred in several fast-moving groups of aquatic animals. At the present time, we have little information on the cost of swimming in marine mammals. The swimming information summarized in Figure 7.4 was derived mostly from fish studies; small fish at that. None weighed over 2 kg. The data obtained from a young sea lion that weighed 30 kg are one exception. At slow swimming speeds of 1.6 m sec^{-1}, its metabolism was about 20 percent higher than when it was resting.

This discussion can be concluded appropriately by considering a common question asked about seals, or for that matter any animal. How much food do they require to maintain themselves? If we assume that a resting, adult Weddell seal requires 300 ml O_2 per kilogram per hour (Table 7.1), we can estimate the minimum value. That value is equivalent to about 650 kcal per hour, or 15,600 kcal per day, if we assume that the burning of 1 liter of oxygen produces 4.8 kcal of heat, an accepted standard value for animals that are utilizing mainly fat and protein for fuel. At the worst, such an assumption will introduce an error of only 10 percent because the amount of oxygen re-

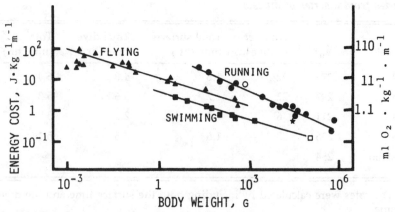

Figure 7.4. The relation between the cost of transport and body size for swimming, flying, and running. O = surface-swimming duck; □ = estimated cost to a por- poise that experiences laminar flow; ★ = young sea lion. (Data from Schmidt-Nielsen, 1972; Goldspink, 1977; and Kruse, 1975.)

quired to produce a given amount of heat is known for all major
foodstuffs (fat = 4.7, carbohydrate = 5.0, or protein = 4.5). (For a
clear discussion of the various fuels and how the energy content was
determined, read the Schmidt-Nielsen work in the Suggested read-
ing at the end of this chapter). According to standard nutritional
texts, herring contains about 150 kcal per 100 g of fish. Assuming that
all this is obtained from the fish, and that the fish eaten by Weddell
seals are equivalent to herring, then 100 fish of 100 g each or 10 kg of
fish are needed daily by an adult, resting Weddell seal. Obviously, a
seal cannot get its fish while resting, nor does it extract all the calories from the fish. I have found at my Scripps laboratory that a rule of
thumb in maintaining young seals is that they will need from 5 to 10
percent of their body weight in fish per day. These are young, growing animals; it should be less for an adult seal. Thus, an adult Wed-

Table 7.4. *Summary of some features of the pulmonary system of
Weddell seals.*[a]

	Weddell seal	Man
Respiration rate (rest)	4	12
(Breaths/min.) (exercise)[b]	15	50
V_T (rest)	8	0.6
(exercise)	12	2.0
LV_I	19	7
LV_D	12	
V_{EDEV}	3.4	
V/M_B (liters/100·kg)	5	7
$\dot{V}_E$ (rest)	30	7
(exercise)	190	200
$\dot{V}_E/M_B$ (liters/100·kg)		
(rest)	7	10
(exercise)	45	285

[a]These are average values taken from 4 adult animals whose mean weight
was 425 kg. All volumes in liters: V_T = tidal volume; LV_I = lung volume after
inspiration; LV_D = diving lung volume; V_{EDEV} = end of the dive expiratory
volume; and $\dot{V}_E$ = ventilation rate in liters/min.
[b]For the seal, this is rate during recovery from a dive.
Source: Kooyman et al., 1971.

dell seal may need for maintenance from 10 to 50 kg of fish daily depending upon its activities. The large amount of fish eaten by the seal mentioned earlier is probably unusual. Those observations were made at the end of the breeding and pupping season. Many of the seals, especially cows that have recently weaned pups, have incurred a great weight loss. December may be a month of intense feeding effort by the seals as they recoup their spring losses.

Finally, in order to give you some idea of the ventilation characteristics necessary to meet the oxygen supply and CO_2 removal needs of the seals after they return from dives, I have compared them to an average man in Table 7.4. As you may note, the tidal volume relative to total lung volume is much larger in the seal than in man. Also, after exercise and prolonged dives, which would cause the seal to ventilate at maximum, the amount of gas exhaled per time ($\dot{V}_E$) goes up only about 6 times, whereas in man $\dot{V}_E$ can increase 25 to 30 times above the resting level. This narrow scope of ventilation is characteristic of marine mammals. They have great capabilities for

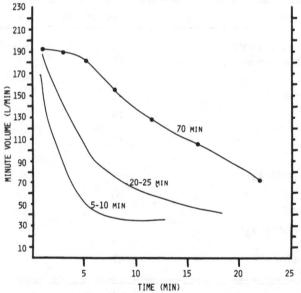

Figure 7.5. The ventilation characteristics of Weddell seals upon returning from dives of different durations. The 5 to 10 min. and 20 to 50 min. curves are average values of numerous measurements. Recoveries from dives of 10 to 20 min. duration were scattered between the two curves. As the 70 min. curve shows, the intensity and duration of recovery for the 70 min. dive was much greater than for any other dives. (From Kooyman et al., 1971.)

breathholding, but poor ability to ventilate to the high degree that terrestrial mammals can. This is perhaps not surprising, considering that their system of respiration is designed to function for extended periods without ventilating, and for intermediary metabolism to function anaerobically for portions of the longer dives. Such dives are interrupted by surface rest periods in order to replenish oxygen stores. The greatest ventilation effort we observed was after a 70 min. dive (Fig. 7.5), one of the longest ever recorded.

Suggested reading

Oritsland, T. 1977. Food consumption of seals in the antarctic pack ice. In *Adaptations within Antarctic Ecosystems*, Proceedings of the 3rd Special Committee on Antarctic Research Symposium on Antarctic Biology, pp. 749–768, ed. G. A. Llano. Washington, D.C.: Smithsonian Institution.
Schmidt-Nielsen, K. 1975. *Animal Physiology: Adaptation and Environment*. Cambridge: Cambridge University Press.

8

Under-ice orientation

(SUMMER DAY – WINTER NIGHT)

> Any error in brachiating from branch to
> branch could be fatal. Every leap was an
> opportunity for evolution.
> *Carl Sagan*, The Dragons of Eden

The problem of under-ice orienta-
tion in Weddell seals is essentially a dual one because of the difference between deep, feeding dives and exploratory dives. Not only is
there a dual behavioral relationship, but because of the great difference between summer and winter environmental conditions the seasonal physical conditions should be considered separately as well.

What physical clues and modes of perception are most important
and how the seal develops its skills both might be better understood
by comparing two types of dives. Navigation to and from the breathing hole during deep dives would seem to be less demanding than
navigation during exploratory dives. The deep dive is of much
shorter duration and distance, and there is always the directional
constant that the hole is somewhere above the seal. Conditions
would seem to be much more rigorous during exploratory dives.
There is no dependable direction to the breathing hole, and the distance from the hole and duration of the dives is much greater.

The sea in which the seals must perform is always changing. A
good time to begin a discussion of the sea conditions in McMurdo
Sound is with the month of October. Not only is it a month when
winter officially ends for the United States Antarctic Research Program (USARP) and support flights begin to bring in fresh personnel
and supplies, but it is also the time that near-term female Weddell
seals arrive in numbers. The coastal margins of the Sound are turned
into a huge maternity ward.

At this early time in the "summer day" when the sun no longer
sets, the entire Sound as far north as Cape Royds is a solid sheet of

ice 1 to 3 m thick. This ice cover is not flawless. Tidal cracks are present along the entire coastline, and there are also other cracks running between islands and out to sea. Some of these have developed randomly as a result of recent strong winds, currents, or sea swell. Most will freeze over within a few days, but others occur consistently in the same places year after year. Although perennial cracks freeze over, the ice is seldom very thick, and the seals can readily break or bore holes through it to breathe and to get out onto the ice surface.

It is still cold enough, and will be through November, so that the ice continues to thicken. There are three discrete parts to the sea ice: The first part is the snow layer that accumulates on the surface. In some areas, no snow accumulates because of strong, unhindered surface winds that polish the ice surface. In others, snow drifts to several feet thick. In snow accumulation areas, the snow's weight pushes the sea ice down and may even cause fracturing. Snow is more opaque than ice, and even thin snowdrifts substantially reduce the amount of light that reaches the water below. The second part is the hard ice that is the rigid foundation floating on the sea surface; the third is the platelet ice that grows on the bottom of the hard ice and gradually merges into it. If currents are not strong, the ice crystals in the water column attach themselves and the platelet ice will grow into large, flat crystals that extend down 2 or 3 m, uniformly covering the undersurface of the foundation ice. There are all sorts of permutations, from no platelet ice layer at all, to patchy or scattered columns and mounds, to platelet layers over 10 m thick. The variety of combinations of different kinds of cracks, added together with snow cover on the surface and platelet ice below, results in a diverse topography as seen from beneath.

As the climate changes, so do the sea ice features. The summer warming has profound effects, and by December the sea ice begins to erode both above and below. Above, the snow and ice melt, forming potholes and melt pools. Below, the currents erode away the ice, and in certain areas large open bodies of water develop. By late January and February, the ice is so weak that strong winds and swells break it up into flow ice. If a strong south storm occurs at this time, the entire McMurdo Sound may be cleared of ice overnight, only to be ice choked the following day if a strong north wind comes up. Until March, McMurdo Sound randomly fluctuates between an open body of water and one choked with ice floes, and occasionally an iceberg. Then, in about late March, the winter freeze begins. Usually by June, the Sound is once again cloaked in a solid sheet of ice.

Concomitant with the changing physical conditions, the biological character of the Sound also alters. After months without sun, the waters of the Sound are perhaps the clearest in the world. Light and dark areas in the ice can be discerned from 300 m, and well-lighted images can be seen more than 150 m away. This exceptional transparency of the water gradually declines as the sunlight stimulates the growth of phytoplankton. Then, in mid-December, as the plankton population reaches a critical level, a true population explosion occurs and the well-known McMurdo Sound "bloom" erupts into a density of living matter so great that the surface water becomes a living soup. Visibility is reduced to 1 or 2 m. However, it is like the dense San Francisco Bay fog that often spills into the Bay and over the land; it is a thin layer. Like the planes that quickly rise above the Bay into brilliantly sunny skies one can dive to 20 or 30 m in the Sound and emerge into clear water again.

Equally as dramatic as the increase in marine microorganisms is the influx of birds and mammals. At the beginning of October, only Weddell seals are present. By the end of the month Adelie penguins are refurbishing last year's nests at Cape Royds, and by the time the ice breaks up in December leopard and crabeater seals, killer and minke whales, and emperor penguins may be seen. All disappear when the winter freeze begins, including all the Weddell seal pups and most of the adults.

The winter night is three months long, and the only light is from the moon, stars, and auroras. During this period, most of the Weddell seals regain the weight lost during the summer's breeding activities. The odd pup or yearling seal that remains over the winter is often not so fortunate. They are usually in poor condition by winter's end. This is also true of the rare leopard or crabeater seal that becomes trapped in this area when the winter freeze prevents it from swimming to the north. Several carcasses and a few emaciated individuals found in early spring are evidence of their misfortune. Such observations further impress one with the uniqueness of the Weddell seal's ability to exist in this area. It seems that the diving abilities of this seal have been honed by the rigorous selective forces of the high-latitude, coastal antarctic environment. The reward is a habitat free of other predators that would compete for the same food sources, or use the seal as a food source.

An important ability that enables the seal to remain in ice-covered seas throughout the year is ice reaming. The head of the Weddell seal is peculiarly shaped in several ways that appear to be specific adaptations for boring breathing holes in ice. First, the head is disproportionately small and narrow relative to the body. This must simplify putting the head in small holes or cracks for breathing in existing holes, for widening them, or for starting new holes in thin ice. Second, a large gape makes it possible to swing the lower jaw clear so

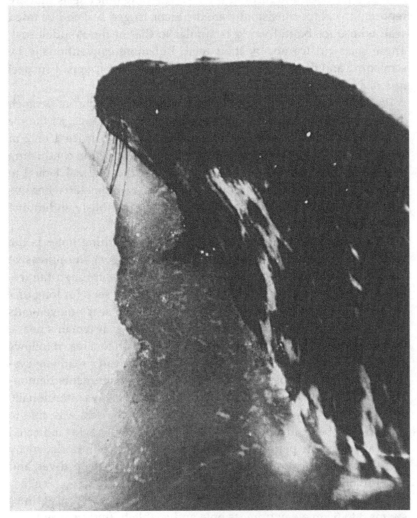

Figure 8.1. Weddell seal cutting ice. The seal makes sideways cuts through about a 90° to 180° arc. Most of the cutting is done by the upper jaw.

that the upper jaw can cut at the flat undersurface of the ice. Third, the upper incisors and canines are exceptionally stout and protrude forward to such an extent that a dentist would recommend orthodontia (see Fig. 2.2). These teeth are the principal cutting tools for ice reaming. With vigorous 90° to 180° arc slashes, the seal abrades the ice with its upper canines and incisors (Fig. 8.1). Sometimes when it is shaping the upper edge of an ice hole, it literally hangs by its teeth as it slashes back and forth at the ice. The only other seals that are reported to cut ice consistently are the arctic ringed seal and bearded seal, whose ice-bound world is similar to that of the Weddell seal. These seals cut ice not by their teeth but more conventionally by scratching and digging with the claws of their foreflippers. I suspect this is a less effective means.

Weddell seals probably do not begin new holes if the ice is much thicker than 5 to 10 cm. Holes are maintained, however, as the ice thickens, so that a breathing hole may be present in ice 1 or 2 m thick. One seal that we kept in water for 3 weeks while conducting respiration experiments cut so much ice that the original 1 by 2 m hole in 2 m thick ice was more nearly 3 m in diameter when we removed the animal. Its effort was beginning to seriously undermine the laboratory.

Equal in importance to its ability to make breathing holes is the seal's ability to find them again. This may not seem so impressive during the summer, when intense light penetrates through the ice, holes, and crack systems. The shaft of light can be seen for long distances in the crystal-clear water. Quite likely, the seal's movements are regulated by the crack systems. While feeding, it remains near a well-marked crack or cracks. When it moves to a new area, it follows leads or perennial cracks that serve as an underwater roadway system. Although in most cases this is true, our experiments demonstrated that this is not necessary. The holes were always intentionally cut well away from any network of cracks, yet the seal was able to find its way back after some very long dives. Its behavior indicated that, after a short period of orientation, it was not intimidated by these isolated surroundings and frequently made deep dives and also prolonged dives.

This kind of navigation requires pinpoint accuracy. A seal at times swims 3 to 6 km away from its hole and is able to find its way back. The short time it has to accomplish this feat because of the duration it can hold its breath leaves little margin to correct for confusion and disorientation. Under the conditions of a summer day, this problem is fascinating. Even more intriguing is the question of whether these skills are diminished in any way during the winter night. To paraphrase Carl Sagan: Every dive is an opportunity for evolution.

There are many possible navigational cues that the seal may rely on in one way or another. The most obvious and likely modes of perception are: tactile, visual, and/or acoustic. For example, there are numerous currents in the Sound that are locality specific. Perhaps the seal may be able to feel them, and, coupled with some compass-heading information, determine approximately where it is.

During the summer, vision must play an important role. The intense light penetrating through the ice gives the appearance of clouds in heavy overcast. At breaks in the ice, the light beams through strongly. In the occasional hole or crack that has no ice floating in it, a bright shaft of light can be seen probing into the depths. In these clear waters, the under-ice surface can probably be seen as deep as 200 m, and diffuse, non-image-forming light even deeper.

Morphologically, the eyes of the seal seem well suited for the task of perception in low illumination. They are large: 60 mm in diameter, compared to man's 24 mm diameter eye. The slit-shaped pupil can open to a large, round light-gathering aperture (Fig. 8.2), and behind

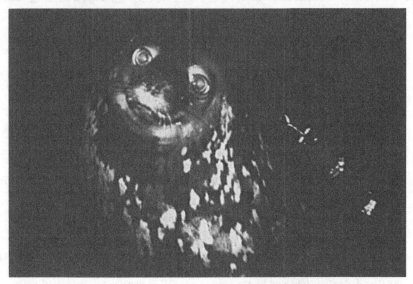

Figure 8.2. An underwater portrait taken in dim light. Note the large, dilated pupils. The seal is carrying a recorder.

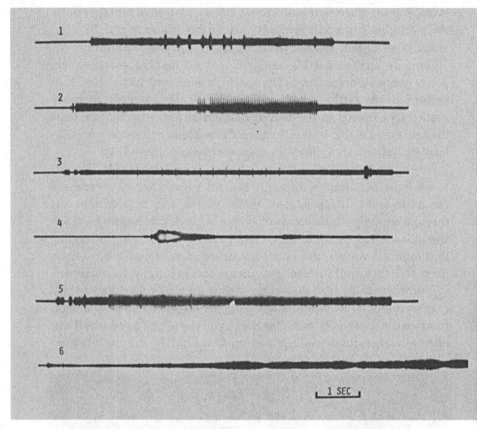

Figure 8.3. (*Above*) An oscillographic representation of the amplitude and duration of several broad categories of Weddell seal calls. The original recordings were obtained from pelagic animals well away from any breeding rookeries. (1) Tweet – indicated by the sharp spikes in a noisy background. Note echo occurring about 0.5 sec after each tweet. (2) Twitter – indicated by the series of sharp peaks beginning midway through a noisy background. The tweet and twitter are the most birdlike calls. These were between 10–15 kHz, and some investigators have recorded similar calls at even higher frequencies. (3) Knock-whistle – the knock is the first deflection, followed by the long whistle. Superimposed on the whistle are the tweets from another animal. (4) A short whistle followed about 2 sec. later by its echo. (5) The steady, low-frequency buzz. (6) One quarter of the record of a long sweep that began at a frequency of about 10 kHz and gradually descended to less than 1 kHz. The rises and falls in amplitude are probably caused by the animal turning its head from side to side, which would vary the strength of the signal reaching the hydrophone. (*Opposite*) Sonagraphic intensity and frequency analysis of vocalizations. Analysis was done on a Sonagraph 6061B 85-16000 Hz spectrum analyzer, from Kay Elemetrics Corp., by Jeff Norris of Hubbs Sea World Research Institute. (A) A tweet with an echo. It is usually not associated with seals near the breathing holes. The plot on the left is the frequency versus time

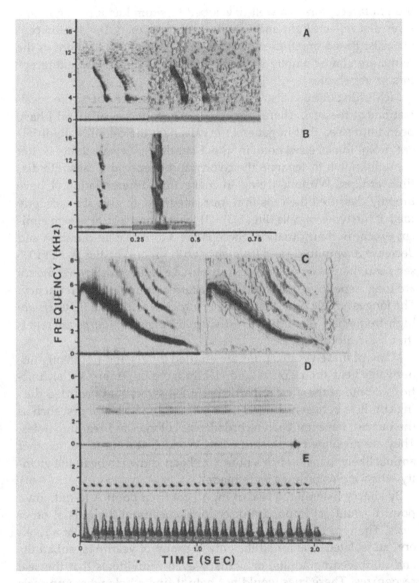

plot. That on the right is the intensity plot. It shows where the greatest energy of the call occurs. Each contour is 6 decibels (db) above background. (B) One of a series of high-frequency chirps. These are made by seals some distance from the hole. The intensity plot shows that there is some energy at frequencies above 16 kHz. (C) A short whistle. This is often heard during interactions between seals near the holes. (D) A long buzz made by a seal near the hole. This call is more often recorded from seals away from the holes. This call lasted 22 sec. (D and E) A frequency analysis of the first and last part of the call, along with (E) an intensity analysis of the last portion of the call (E).

the photoreceptors is a well-developed tapetum lucidum, a guanine layer that reflects light and enhances stimulation of the photoreceptor cells. Based on studies of other seals, the photoreceptors in the retina are almost wholly rods, the most light-sensitive photoreceptors of vertebrates.

McMurdo Sound is also a noisy place underwater due to the vocalizations of the seal. There is much variety to these calls, and I have been impressed that in general the calls near the colonies are different from those elsewhere in the Sound. However, there is not enough known to separate the colony and deep-water calls into distinct groups. Without trying to make this differentiation, I have broadly classified the calls into four categories to give some orderly idea of the types of calls (Fig. 8.3). (1) *Sweeps* and *whistles* seem similar, except in their duration. Sweeps are 3 to 20 sec. in duration and decrease in amplitude and frequency with time. Whistles last 1 to 1.5 sec., and there is a rapid drop in amplitude and frequency. (2) *Buzzes* are long, repetitive emissions of constant amplitude and frequency. The longest single buzz that I have recorded was 22 sec. (3) *Tweets* are high-frequency, repetitive, short-slurred bursts. (4) *Chirps* are repetitive, staccato bursts.

Many of these vocalizations are emitted loudly and frequently underwater near the colonies and clearly have important social functions. Some of these calls, particularly the sweeps, are emitted during conflicts between animals over a breathing hole. Others, such as the buzzes, have not been recorded near colonies or breathing holes. They are presumed to be emitted by Weddell seals because no other animal likely to make such a noise has been known to be in the vicinity when the recordings were made.

By simply listening, a seal trying to get from point A to unknown point B could get important directional cues from the calls of other seals. The type of vocalization might tell it whether B was at a rookery, an isolated hole for adults only, a group of yearlings, subadults and nonbreeding adults, or perhaps even individuals that the seal recognizes. These cues would not help it find a hole where no seals were present or were not vocalizing. Passive listening could be dangerously misleading if the searching seal swam toward an animal that was some distance from a breathing hole.

A more positive method of using sound would be for the seal to generate its own navigational cues by echo ranging. One of the seals with which we were doing respiration experiments occasionally would make some sounds that are tempting to speculate about. The sounds were made just below the laboratory and were so powerful that we could hear them through the ice and hut floor. Later, they were recorded with a hydrophone, and a distinct echo can be heard. Analysis of the call shows that it could easily function as a depth-sounding signal (Fig. 8.4). The pulses begin well spaced and then are

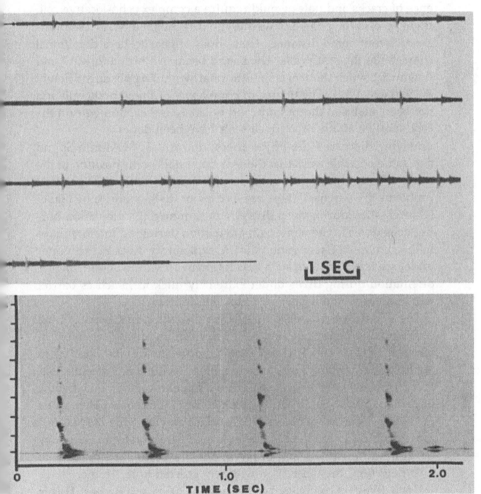

Figure 8.4. (*Above*) An oscillograph show-ing amplitude and duration of a series of chirps (large deflection) and the echo (small deflection). This call was recorded from an experimental seal in November. The sounds were produced while the ani-mal was underwater, a short distance from the breathing hole. (*Below*) Sona-graphic intensity and frequency analysis of a similar call, except no echoes show and the pulse interval lengthened after each pulse, rather than shortened.

speeded up until another pulse is sent out before the echo returns from the first. Each series consists of 15 to 17 pulses. Such a vocalization might give information on depth, presence of a submerged cliff face, or other major bottom features. Why the animal changes the repetition rate of the pulses is not understood.

More precise information about the under-ice surface or the presence of cracks and holes would require a complex echo-locative call. It would also have to be powerful and directional to yield useful information at some distance. There does appear to be a directional character to the seal's calls. The sound beam is directed forward and downward when the seal is in its normal swimming posture (Schevill & Watkins, 1971). The degree of complexity of the various calls has not been analyzed thoroughly, and no analytical experiments on the echo-locative ability of Weddell seals have been done.

Having discussed the physical environment of McMurdo Sound and some possible ways that the seal may establish its position in the water column, let us now consider the specific kinds of orientation problems the seal has. There are five major distinctions to be made: (1) orientation during deep dives in the summer; (2) orientation during deep dives in the winter; (3) navigation during exploratory dives in the summer; (4) navigation during exploratory dives in the winter under sea ice of 2 to 4 m, such as is found in McMurdo Sound; and (5) orientation during dives under extremely thick ice, such as barrier ice, in areas where it may be 10 to 300 m thick.

The deep dives, based on evidence presented in Chapter 5, are considered to be for catching fish. They can be further separated in terms of orientation: (1) The seal must be able to find its way back to the hole after random turns and diversions as it pursues fleeing fish, and (2) the seal must be able to perceive somehow fish in inky depths of perhaps 600 m. Because these are two very different problems, one of finding a stationary object at the surface and the other of finding a moving object in the depths, the nature of detection for the two may be quite different. As my observations are pertinent primarily to hole detection, the discussion will be restricted to this topic.

Deep dives are considered to be those of 200 m and deeper. This is not to say that the seals do not catch fish at shallower depths, but rather that such depths are not considered a difficult orientation problem during the summer.

The usual duration of deep dives is 8 to 15 min., and they rarely exceed 25 to 30 min. For purposes of some simple calculations, let us assume a 15 min. dive. The departure and return are almost always perpendicular to the surface, and the departure is with vigor. Assuming the average swimming speed throughout the dive is 10 km per hr., then in 7 to 10 min. the seal could be 1,000 m from the hole. Four hundred meters of this distance would be directly below the hole, and the rest would be used up in random tracks as the animal pursues fish. These pursuits could take the animal a horizontal distance away – as much as 900 m. If at this time the seal starts up, this one course change of swimming upward insures that the seal will get closer to its goal. The greatest distance it will be away from the starting point when it ascends to near the under-ice surface will be 900 m. This distance is unlikely to be beyond the area that is familiar to it. When the seal reaches 200 m, and perhaps deeper, it becomes possible to visually scan the ice and assess its position. If echo ranging is employed, it may be able to begin determining where under the ice it is at even greater depths. Thus, it does not seem that summer deep diving presents a difficult orientation problem. Winter dives, however, would seem to be a much greater problem if vision plays an important role. Without knowing how sensitive the seal's eyes are, it is impossible to know at what depth the animal begins to see landmark features of the ice. Evidence I will discuss later provides some information on the matter. If the seal has echo-locative abilities, they would be especially helpful at this time of year.

Orientation during exploratory dives is different, more difficult, and a higher risk to the seal than during deep dives. These extended forays are often into unfamiliar areas. Unlike deep dives, there is never one directional component that is a constant toward the starting point. As soon as a seal makes a turn, or even gradually drifts off heading, there is the potential for utter confusion because the hole is no longer directly behind. This is particularly true when the seal is diving from the laboratory, because it is then at an isolated site, whereas the natural breathing holes are usually part of a system of cracks with openings spaced along the crack. Yet, over years of studying seals diving from the station, the impression is given that the seals are not often confused, and never fatally lost, even though sometimes these dives take them several kilometers away from their starting points.

The under-ice chamber observations have provided one clue. Almost invariably, in 90 percent of 85 observations, seals returned from the same direction from which they had departed. My conclusion about this behavior is that the seals maintain a constant heading during exploratory dives and do not deviate. This then gives them a reliable return heading to their starting point. If their swimming speed is held constant and their timing is good, then they know accurately where the breathing hole is. How do they hold a steady heading? Any scuba diver will confirm that this is not easy to do; in fact, it is impossible unless there is some bottom feature, such as sand ripples, that has a constant directional attitude. For the seals, the question remains unanswered.

Not only did the observations on headings show a strong relationship between the directions from which the seal would depart and return, but there was a bias to certain compass headings. These biases were toward the nearest shoreline. How can a seal diving under ice with visibilities of 100 to 200 m tell where a shore several kilometers away is? The near-shore area is almost always one of seal concentrations, and, most logically, during the summer months when these observations were made the seal heard the calls of others.

This is not the only cue the seals may use. During one summer, we established an exceptionally remote experimental hole, and the seals were unable to escape from it over a period of several days. Because it was necessary to get rid of one seal before beginning another experiment with a fresh animal, I had the perplexing problem of how to get a 400 kg animal out of the ice hole. The problem was solved for me one day when I left the hut door open while working outside. I noted that the seal took great interest in looking out the door, which happened to be facing the shoreline. The coast has many prominent features and is dominated by 4,000 m high Mount Erebus, a volcanic cone. Shortly afterward, the seal disappeared forever.

After each experiment was complete, I tried this again; in each case, the seal would rise well out of the water and above the floor of the hut and look intently out of doors. And, in each case, four in all, the seals soon disappeared. Furthermore, in one instance a seal escaped from the laboratory by swimming to another equally remote but natural breathing hole ("natural" in the sense that the seal had found a thin spot and taken the time to bore an escape route to the surface). After it hauled out, the hole froze over and it decided to go across the ice. It headed directly toward Mount Erebus. In this situation, there is no possibility of acoustic orientation. The conclusion is that the seals do use visual landmarks on the surface, and these aid them in their underwater traverses.

That the sled in which the seals were transported was not covered may explain in part some exceptional first-dive efforts. On three occasions, seals released at the laboratory escaped on their first diving effort. The evidence obtained from these dives after the instruments were recovered tells how far the seals are willing to go. The longest recording of the three occasions extended to 50 min. at which time the animal was at a depth of 70 m (Fig. 8.5). Unfortunately, the timer stopped at this time, and the total length of the dive is unknown. The distance swam must have been at least 10 km. During the next-longest recording, the timer stopped at 30 min. when the animal was at a depth of 110 m and still descending. The shortest of the three occasions lasted 26 min. (Fig. 5.3).

This brings us to another important point. The durations of exploratory dives range from 20 min. to at least 73 min., but most are between 20 and 40 min. The calculations of Chapter 6 tell us that even a 73 min. dive has at least a 20 percent margin of safety. Only infrequently does a seal exceed even 50 percent of its breathholding capacity. The few occasions that one does (i.e., dives longer than 60 min.) may represent the few times that a seal has become confused and taken longer to return than it planned.

Numerous recordings of time-versus-depth profiles have been obtained in the summer; these are informative. Few dives that exceed 30 min. also exceed 200 m. The average maximum depth of exploratory dives is 130 m, and, again, most of the dive is at a lesser depth. Even at the maximum depths, a seal can see the surface.

Surface scanning is simplified in the Weddell seal by the mobility of its eyes. They are able to roll 90° in their sockets. This permits

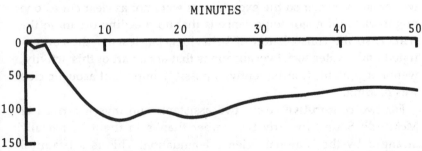

Figure 8.5. Time–depth profile of the first dive of a pregnant female after release into the hut ice hole. After 2 min. she blew bubbles up into the hole, but never surfaced. She was later found at the rookery where she was originally collected. (From Elsner, Kooyman, and Drabek, 1970.)

viewing directly above or forward without moving the head and altering the most favorable hydrodynamic body conformation.

Finally, hydrophone recordings at the point of origin of the dives did not often yield information indicating any unusual vocalizations by the experimental seals. However, the hydrophone was omnidirectional, and it was presumed the calls of the known seals would be louder because of assumed greater nearness to the hydrophone. This kind of acoustic information is highly circumstantial and is considered marginally useful.

A review of the major features of the Weddell seal diving during the summer months is as follows:

1. Deep dives have these characteristics: (a) short duration (5 to 25 min.); (b) short distance from the diving hole (less than 1,000 m); (c) always below familiar under-ice topography; (d) one component of the starting point is constant (it is toward the surface); and (e) the presumed working depth of 200 to 400 m. is in a region of low ambient light levels.

2. Exploratory dives have these characteristics: (a) long duration (20 to 73 min.), but rarely approaching a breathhold limit of the seal; (b) long distance from starting point (sometimes greater than 12 km); (c) often below unfamiliar ice; (d) diving depth does not exceed the limit of visual contact with the surface; and (e) the point of origin is held at a constant heading by the seal.

It seemed to me that crucial to learning more about the nature of the Weddell seal's method of orientation was to dive animals at night when vision would be greatly restricted. My hypothesis was that if vision was important the diving behavior of the animals would be altered. Unfortunately, other variables were not constant between winter and summer so the experiments were not as clear cut as one might wish. Of major importance is that no breeding occurs in the winter, so the constellation of behaviors, particularly the concentrated underwater social vocalizations that are a part of this activity, were not present. Consequently, a possibly important acoustic clue was missing.

For two consecutive years, two assistants and myself arrived at McMurdo Sound in early September thanks to flights especially arranged by the National Science Foundation. This is a beautiful month at McMurdo. Much of the time there is no wind, and on clear nights the Milky Way cuts a brilliant path across the sky. I have never seen brighter stars anywhere. Exceeding this spectacle are the hours-long, glowing gold sunrises, sunsets, moonrises, and moonsets. It is also a month of rapid change from winter to summer, for the sun is rising in the sky at its maximum rate. In the beginning, the sun is above the horizon six and a half hours; by the end of the month, it is there for sixteen and a half.

From the beginning, I began to note contrasts in the seals' behavior from the summertime. During the initial week of our first season's work, I became quite concerned because we could not find any seals even though we drove considerable distances and inspected traditional haul-out areas in our search. There was no question that seals were present. There were active breathing holes, unfrozen and recently reamed, and with our hydrophone we could hear underwater vocalizations. However, both the number of holes and seal calls were much fewer than during the summer.

After one discouraging day, when I was beginning to have doubts about the worth of our long journey to the Antarctic for this study, we saw a seal haul out in the dim late afternoon light. We forthwith captured the animal, and while we were doing this two more hauled out nearby. This was interesting because during the summer the seals increase in number all "day" (0400 to 1600 hours), peak at 1600 hours, and then steadily decline until they are at a minimum by 0300 to 0400. During 1964, in a nearby area, 450 seals were counted at 1600 and only 20 at 0300 (Fig. 8.6).

This prompted us to do surveys during September, when weather permitted. In 11 nights of counting, 48 animals were noted; during 12 days of counting, only 16 animals were seen (Table 8.1). These

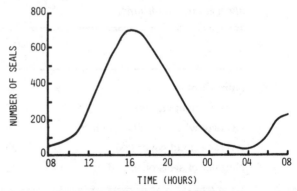

Figure 8.6. Daily rhythm of seals lying on the ice just south of Cape Armitage on February 17, 1963. (From Smith, 1965.)

differences are more remarkable than the numbers indicate because in the day the entire area in which the survey was conducted can be assessed from the centrally located, 30 m high Turtle Rock. At night, the seals had to be found individually by locating them in the head-lights of the survey truck and assuredly some were missed. In Sep-tember, the seals have at least altered their haul-out behavior and are out during the night hours rather than during the day.

Once we discovered the secret to finding seals, three experiments were conducted during each season. All were adult animals, three were males, and the body weights ranged from 360 to 450 kg (Kooy-man, 1975).

A few comments about the procedures: At all times, a 1.2 m high blind was around the hole so that our activities in the hut would not disturb the seal. At night, the only light was a flashlight and reflector directly over the observer's notes. While the seal was present, con-versations were only in hushed tones. Under these circumstances, the seal's activities were monitored continuously at night. This was also done as much as possible in the daytime, but at times it was necessary to leave the seal untended. In all, 215 hr. of direct observa-tions were made. In addition, time-depth recorders (TDRs) were of-ten on the animals and these added many more hours of observation for which we knew at least the maximum dives the seals were achiev-ing.

Table 8.1. *Number of seals observed on the ice near Turtle Rock, Mc-Murdo Sound, Antarctica, during the day and night, with the prevail-ing weather conditions.*

	Day (1600 hr.)	Night (2100–2330 hr.)
Number of counts	12	11
Average temperature (°C)	−19	−16
Average wind velocity (km/hr)[a]	20	25
Average cloud cover (%)	20	66
Number of seals observed	16	48

[a]The wind velocities were estimated by the observers, and thus are crude values. The observations were made between September 6 and October 1, 1971.

Because of the long sunrises and sunsets, it seemed worthwhile to separate some of the data into day, night, and twilight observations. Twilight was considered 3.5 hr. before sunrise and after sunset.

From the first experiment, it appeared that there was a difference between the seals' behavior in the daytime and at night. The seals spent an average of 58 percent of their time in the day diving and 44 percent at night. Our impressions were that the differences were even greater than the numbers told us. Our subjective impressions were enhanced by the greater vigor of the daytime dives, which a simple statistical analysis of time above and below the surface does not give.

Depth-of-dive measurements gave us solid evidence of the difference. The seals reached depths as great as 600 m during the day, but never exceeded 270 m at night. However, one could argue that this difference is not due to any alteration in the seals' enthusiasm for diving, but rather that the animals they feed on are in shallower water at night. It is a well-known phenomenon that midwater animals have a diurnal cycle of depth distribution. At night, they come to or near the surface; in the daytime, they descend. Yet many of the night

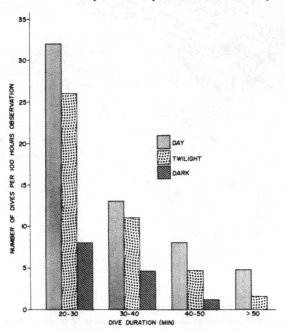

Figure 8.7. Frequency of exploratory dives during September. Twilight dives were those that began between sunset and 3.5 hours after, or between 3.5 hours before and until sunrise. (From Kooyman, 1975.)

dives did not seem to be for feeding, but rather were just short, loaf-
ing dives. Furthermore, much of the seals' time on the surface at
night was spent sleeping, whereas during the day surface time was
used for recovery and preparation for more dives. However, our
most crucial clue was the differences in exploratory dives.

Fewer exploratory dives were made at night or in the twilight
hours than in the day (Fig. 8.7). The difference ranged from 2.8 times
more 30 to 40 min. dives, to 6.6 times more 40 to 50 min. dives in the
day than at night. No dives greater than 50 min. were observed at
night.

The most dramatic differences of all, and to me the most informative, are the time-versus-depth profiles of exploratory dives. And
this is where an awareness of the difference between twilight and
night conditions is most important. In the twilight hours, the sky is

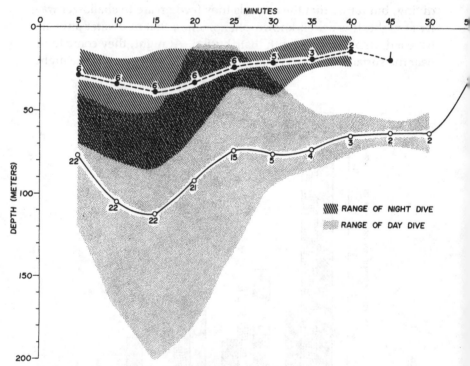

Figure 8.8. A comparison of exploratory
dives obtained between sunset and sunrise in September, and those obtained in
October and November, when the sun
does not set. Closed circles are the means
of after-sunset dives and open circles are
the means of dives made when the sun
was above the horizon. Numbers next to
the circles are the sample sizes. (From
Kooyman, 1975.)

still bright and silhouettes of objects are distinct. Under-ice configurations, especially cracks and holes, are discernible. Consistent with these low light levels, profiles of twilight exploratory dives are much shallower than any recorded in the day (Fig. 8.8). Even the deepest point in all these twilight dives and the one night dive observed do not reach the mean depth of the day dives.

My conclusions from these various bits of information are:

1. Under our experimental conditions, seals preferred diving in the day to night.
2. Even under natural conditions, there is a preference for daytime diving, as indicated in the difference in the diurnal haul-out behavior in September as compared to midsummer.
3. The seals prefer to make deep dives during the day.
4. Most exploratory dives occur during the day when the seals can maximize their visual cues.
5. The shallowness of night exploratory dives occurs because of one or more of the following:
   a. In such dim light the under-ice surface is visible only at shallow depths.
   b. If seals have echo-locative ability, it is good for orientation purposes only at short distances.
   c. If the seals are using both vision and acoustics for orientation under night conditions, the sum of information from both systems is optimized at shallow depths.

In summary, I believe that all the evidence indicates that vision plays an important, if not the dominant, role in seals' under-ice exploratory activities.

How is such a conclusion compatible with the fact that for nearly three months of winter the seals that remain in McMurdo Sound have no choice but to dive at night? The sun sets in late April and does not rise again until mid-August (April 29 to August 15 at the McMurdo Weather Station). During this time, they fatten up from their summer losses. Much of the answer may be in the difference between the physical conditions and chronology of events for a seal diving naturally and those for a seal forced to dive at our artificial site. Those seals that winter over are there at the beginning of ice formation, which usually occurs before the winter nightfall. In the area where they dwell, there are numerous breathing holes, tidal cracks, and perennial cracks. These are all learned well by the seal. From these zones of surface access, the animals swim back and forth, down and up after fish. All their activity is confined to a well-known area and to deep dives. Close inspection of a small area of under-ice surface is enough for them to determine the direction to a breathing hole. Very likely, few or no exploratory dives are made because any long-distance movement is along a crack system of closely spaced breathing holes. Conceivably, no working dives of more than 15 or 20 min. duration are made during the entire winter. This is less than 20 percent of their breathhold capacity. The seal may find new areas where the ice is thin enough to be penetrated. If so, new breathing holes can be made and a new region can be exploited. If the area is sufficiently isolated and the seal is aggressive toward intruders, it may be able to keep the area to itself for an extended period. In this perspective, an exceptional breathhold capacity is a special advantage to a seal diving under ice – providing it with a wide margin of safety during feeding dives if it should become lost, enabling it to search extensively under the ice for new feeding areas, and giving it time to cut away at the ice before returning elsewhere for a breath.

(The above statements were written prior to my most recent investigation of free-ranging seals. (Kooyman and Castellini, unpub. observ.) In that study, we found that dives in excess of 30 min. are extremely rare. Once again, we also confirmed that seals in this area prefer to dive in the daylight hours; see Fig. 5.5.)

All that has been discussed must be tempered with the fifth condition – diving under barrier and shelf ice (ice that flows out from the continent, does not go out annually, and is very thick). There is a small group (perhaps 30) of Weddell seals living year round at White Island (Fig. 8.9). How these animals originally established themselves is unknown, but they have been there ever since the modern antarctic program began in the late 1950s. Each year, two or three pups are born. Barrier ice separates the island from the annual sea ice 20 km away. Near the northern shoreline of White Island, there is a thin area due to the shearing forces of the Ross Ice Shelf as it flows past White Island. On the south side, ice piles up against the island. At the eastern tip, there are tremendous shear cracks 15 to 20 m deep and large pressure ridges 5 to 10 m high. The ice here must be 100 m or more thick except in the cracks.

Among the pressure ridges, tidal cracks, and shear cracks of the northern and eastern parts of the island, the seals find access to and from the water. Platelet ice jarred free from below chokes the crack systems, so that in some places the seals must burrow through these crystals to reach the water.

In 1978, a New Zealand group had cut a 1 m diameter hole through a thin section of ice on the north shore. They were conducting oceanographic studies by routinely collecting samples from the water column. At least three seals were using this hole from time to time. I elected to take advantage of their tremendous labor in cutting this hole. So Dan Costa, a postdoctoral student, and I made a dive down the hole to get a firsthand impression of what the diving conditions were for the seals. It was an ideal day, brilliantly sunny and warm, and the time was noon. We were not psychologically prepared for what was to come. It was one of the most unnerving dives I have made in ten years of antarctic diving.

About 2 m of snow was on the ice surface, the sea ice was 4 m thick, and the sea bottom was at 75 m. The thickness of the platelet ice was unknown, but we assumed it to be about 5 m. Platelet ice is the greatest hazard because it is unstable. Any contact with it causes pieces to break loose and float to the surface. If too much of this dislodged ice collects, it is impossible to swim through it. Because of this, we used single tanks rather than our usual twin-tank packs, in order to reduce the amount of ice we dislodged as we descended.

I led, working my way slowly down a 25 m weighted line. I noted

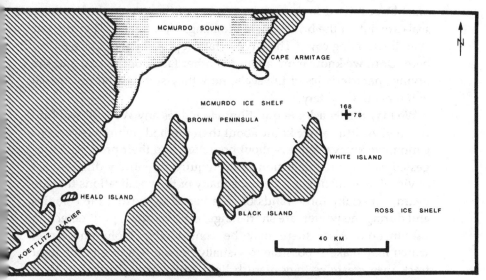

Figure 8.9. White Island is located about 22 km south of the edge of the McMurdo ice shelf. North of this boundary, the ice usually breaks up and is reformed again on an annual basis. The arrows indicate where the seals normally haul out. The two most northern arrows are the places where the three mother/pup pairs were found in 1978. The cross is the location of the 78° parallel and 168° meridian.

as I passed the bottom of the sea ice that the platelet ice looked un-usually compressed; I would soon discover why. The hole narrowed as I descended, and the available light diminished rapidly. At 15 m, I still was not clear of the platelet ice! No wonder the shallower sections of platelet ice were compressed; the upward force on the crystals must be great. Soon thereafter, I signaled Costa to return to the surface because there was no advantage of a second diver in these close quarters. Also, I knew that a lot of platelet ice was being dislodged by the two of us. At 20 m, I was clear of the ice and at the end of my rope. I would go no further and I was concerned that a seal might go into the hole and block my return to the surface, or be startled and dislodge so much ice that my return would be difficult. My time below the ice was short for all these reasons. I noted that it was incredibly dark and I could not see anything out in the water column, nor could I make out any portion of the ice beyond the hole, except what the beam of my flashlight illuminated.

After surveying the under-ice conditions at White Island, I am no longer impressed by the orientation capabilities of Weddell seals – I am in awe. All that I have seen and noted in McMurdo Sound is child's play compared to shelf ice conditions. No marine mammal exists under more rigorous conditions than these. Their movements must be restricted to the nearby regions of the tide crack. Yet these seals are not on the brink where life hangs by a thread. They are fat, even the nursing cows! The largest, fattest pups I have ever seen are here. Until we know more about the other fauna of the area and the diving characteristics of the seals, how they survive and do so well will remain a mystery.

We may finally ask the question: How does any Weddell seal find its prey? As little as we know about their methods of under-ice navigation, we know even less about how they find their prey. As I suggested earlier, prey detection could require an entirely different set of visual or acoustic cues. Perhaps many of the vocalizations that are heard, especially those kinds of calls heard away from the colonies and during the winter, are short-range, echo-locative, prey-detecting in function. Also, there may be considerable bioluminescence, which may make it possible to visually pursue fish due to the light track they create as they disturb light-emitting organisms in their path. Until we improve our experimental techniques, this problem will remain an enigma.

9

Distribution, abundance, and mortality

Out, out brief candle
Life's but a walking shadow.
Shakespeare, Macbeth

Occasionally, Weddell seals have been found north of the antarctic convergence (Erickson & Hofman, 1974). The most northerly record is a sighting off Uruguay at a latitude of 35° S. Several other sightings have been reported in various parts of South America and New Zealand. There is one record for Australia (Fig. 9.1). These are all wandering animals, and they do not represent established or incipient breeding colonies.

The most northerly breeding colony is just south of the antarctic convergence in Larsen Harbor, at the southern tip of South Georgia Island. In this colony, between 25 and 30 pups are born between August and September each year. The total population in the harbor area is perhaps 85 animals. There are only occasional sightings elsewhere on the island.

Weddell seals are present constantly around the entire coastline of Antarctica. At suitable places throughout this area, breeding colonies may be found at the appropriate time of year. Some of the largest concentrations of Weddell seals have been noted in the Weddell Sea, where one survey recorded from 15 to almost 35 animals per square km.

The best census data and natural history observations have been obtained near research stations. For example, some of the earliest work was done by Alton Lindsey in the Bay of Whales. Lindsey noted as many as 500 seals in the area. In Adelie Land near the French base, Dumont D'Urville, a few hundred seals reside. There are 150 to 200 seals born annually near the Vestfold Hills, a site close to Mawson Station, the permanent Australian base. Similar colonies are no doubt present near other parts of the continent.

The Vestfold Hills, which are at a latitude of 68° 35′ S, are an interesting comparison in some ways to McMurdo Sound. At the more northerly latitude, where the coastline is more openly exposed to the Antarctic Ocean, the seals disperse after the pups wean and there is no buildup of seals through the summer as there is in the southern portion of McMurdo Sound. Also there are no breathing holes established in the winter because there is much open water access around icebergs and tidal cracks. Perhaps this abundance of open water represents the more usual environment for Weddell seals, and the common dispersal pattern is like that near the Vestfold Hills. Much less has been said about these populations than the much more extensively studied seals of McMurdo Sound. Most of the following remarks on seasonal changes in population size and dispersion are drawn from McMurdo Sound, one of the only places of which I have firsthand knowledge; it should be kept in mind that McMurdo may be the exception rather than the rule.

In 1902, the largest pupping colony in the whole area from Cape

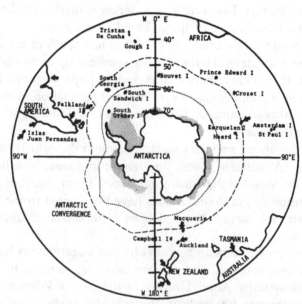

Figure 9.1. Unusual occurrences of Weddell seals in the southern ocean (arrows). The most northerly breeding colony is at South Georgia. The general location of the antarctic convergence is depicted (dashed line). General representations of the northern limits of the antarctic pack ice in winter (dotted line) and summer are shown (stipled area).

Royds to Pram Point was at Pram Point, (Wilson, 1907) (Fig. 9.2). (I
will refer to these localities often in the following discussion, so the
reader might want to become familiar with the map.) This location,
being convenient to the Hut Point base of the 1901–4 British Antarctic Expedition, was the main source of their dog food. In 1959, when
a permanent base was established at Hut Point, 12 pups were born at
Pram Point (Stirling, 1971a). In my most recent observation (1977), 1
pup was born there. This change in the number of pups born illus-
trates the fact that, for unknown reasons, the pupping areas do
change.

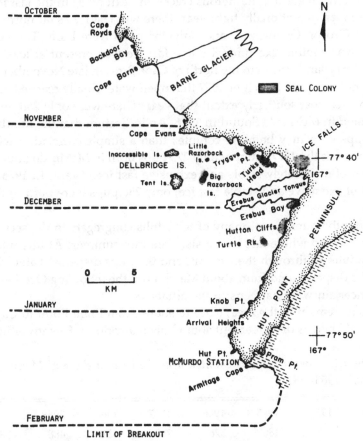

Figure 9.2. The present distribution of
seal-breeding colonies in McMurdo
Sound, giving an estimation of the ice-
front location in McMurdo Sound during
the summer. This is highly variable from
year to year, and the indicated location
and date of the ice front is only to give a
coarse impression of how it changes
through the summer.

Large changes in the number of pups born in McMurdo Sound have been recorded in the last 20 years (Table 9.1). In 1958, there were 5 pups at Pram Point, fewer than 30 at the Dellbridge Islands, Cape Evans, Cape Barne, and Cape Royds; and none at Turtle Rock (Stirling, 1971a). Unfortunately, no information is available on the numbers at Turk's Head and Hutton Cliffs, but apparently there were very few pups born that year in McMurdo Sound.

From 1963 to 1967, there was a steady increase in the number of pups born between Cape Royds and Pram Point. In 1963, there were 380 pups counted; this rose to 767 in 1967. It was noted that 1967 was exceptional regarding numerous cracks and exit holes in the fast ice. In 1968, an exceptionally tight year, there was fast ice as far north as Cape Bird on October 15, and only 456 pups were born. The year 1976 was another exceptional one – fast ice was present at least to Beaufort Island, 16 km to the north of Cape Bird, in late November. It was not until late March of 1977 that open water finally reached Hut Point, its most southerly extent that year. There were only 288 pups counted in McMurdo Sound in 1976. An indication that the number of pups born may be more complex than a simple correlation with fast ice was the 1977 count of only 273 pups born in McMurdo Sound; a year of moderately stable and extensive fast ice. Again, in 1978, a year of extensive and stable fast ice, only 250 pups were born in the area.

After the pups wean, many of the adults congregate in the sectors of the Sound where fast ice persists late into summer. At this time, the adults go through their moult, and the pups disperse. Later, the adults disperse, and from about March until the following October it is uncertain what happens to the adults.

Some remain in the area, close to shore, from Pram Point to Cape Royds. There is disagreement about how many do so. Few investiga-

Table 9.1. *Number of pups counted near the eastern shore of McMurdo Sound,[a] 1963–78.*

1963	1964	1965	1966	1967	1968	1969	1970
380	401	480	578	767	421	488	585
1971	1972	1973	1974	1975	1976	1977	1978
434	501	558	464	—	288	274	250

[a]Area surveyed was from Cape Royds to Pram Point.
Sources: Siniff et al., 1977; Thomas et al., 1978.

tors have overwintered, so specific attempts to assess the population from March through October are not available.

Edward Wilson observed in 1902 and 1903 that many seals were present, and he felt the entire population of adults remained local (Wilson, 1907). However, those years were perhaps exceptional because open water was present throughout the winter only a few kilometers north of Hut Point.

In 1963, Murray Smith, at the time a graduate student at Canterbury University, New Zealand, overwintered. One of his objectives was to take a census of seals during the winter. From July to September, he made four single-day journeys from Scott Base at Pram Point to Cape Evans. On these winter trips, he counted 150 breathing holes (Smith, 1966). In late winter aerial censuses on September 7 and October 10, 1963, he counted respectively 280 and 225 seals, which suggested to him that 250 seals wintered in McMurdo Sound. Based on the limited number of censuses and the fact that many seals must have been in the water during the aerial surveys, this conclusion is unwarranted. It also seems to me that 150 breathing holes for 250 seals is a large ratio. Commonly, in October 2 to 3 seals will be hauled out around a single hole, and a few areas will have at times between 5 and 10 animals.

Based upon my own observations made in two consecutive years in the month of September, I believe that there are not nearly as many seals in the Sound during the winter as in the summer. For example, few seals are to be seen during under ice searches in September in traditional breeding areas. Certainly the seals are dispersed, but the extent is unknown, so we shall speculate some. If fast ice is the preferred habitat of Weddell seals, then as the sea ice freezes during the winter seals are able to disperse considerably. I suspect that this is what the adults do. The year's crop of pups is a different matter, and I will comment on them shortly.

In early October, seals begin to appear in numbers around the pupping and traditional haul-out areas. The majority of these are probably pregnant cows, but males and many nonparturient cows are among them. There is a continual increase of seals found resting on the ice through the summer as more males, nonpregnant cows, and immature animals appear. The aggregations of seals on the ice become denser as the ice breaks up and the fast ice areas shrink. By late summer, there is a considerable concentration of seals around the base and along the edges of Erebus glacier tongue, next to the barrier ice near Pram Point, and along the barrier–sea ice junction that extends from Pram Point well out into the Sound (Fig. 9.2). Many seals are here because these are the only haul-out areas around. Whether they remain nearby while feeding is not known, but it is likely because there is a consistent daily cycle of seals on the ice and the same seals are seen at the same haul outs from day to day (Fig. 8.6). It should be noted that such large numbers of seals as Smith noted in late summer have not occurred in recent times near Pram Point. On February 10, 1978, at 1745 hours, a peak time and an ideal day, only 217 seals were seen (fewer than a third of what Smith counted). On January 20, 1979, at 1830, on a similarly ideal day, only 174 seals were counted.

From February until October, little is known about the dispersal of the seals. The summer expeditionary season is over and everyone has gone home.

Some of the dispersal of the seals may be by age groups. For example, as I mentioned in Chapter 3, on January 7, 1965, about 75 km northwest of McMurdo Station and near shore to the antarctic continent I saw a group of between 150 and 200 seals collected about an ice-locked and perhaps grounded iceberg (Kooyman, 1968). About 70 percent of these animals were subadults, yearlings, and pups. From this group to Dunlop Island another 400 animals were seen from the helicopter, and most of these looked like subadults or pups.

At the time I made these observations, not much significance was attached to them. However, in years of marking and attempting to relocate tagged seals other investigators have found that pups, yearlings, and even two-year-olds are quite uncommon. When they are found it is usually in middle to late summer, when the sea ice is deteriorating and there are many breathing holes and cracks present.

This poses a mystery. Where do the pups go? Is there such a high mortality among young animals that few pups are likely to be seen again in subsequent years? Are the returns poor because the pups, in later years, select other areas to spend their summers? Or are the returns poor because the marking procedures are inadequate for identifying individuals?

Considering all these unanswered questions and the lack of data for winter behavior of seals, it is bold to generalize about the mechanisms of winter distribution. I will do so as a means of summary, and at the risk of drawing criticism from my ecologist friends who see the flaws in my hypothesis.

Let us assume that during an average year the summer peak adult population from Cape Royds to Cape Armitage is 2,000 animals. Let us also assume that in winter there are 200 breathing holes along the reliable crack systems in both the area that Smith probably surveyed, and also in the area several kilometers offshore that Smith probably did not survey. That would mean about 10 animals for each hole if all the adults remain in the area. Based on my own observations while working at isolated holes offshore, the seals are very antagonistic toward one another when near the breathing holes. The vocalizations heard throughout the spring and summer suggest aggressive behavior among seals in widely scattered parts of the Sound. Presumably this aggression is centered around breathing holes. This strife alone would be great incentive for pups, subadults, and weak adults to clear out and go where the sea ice is more fragmented and the breathing easier. Thus, by this hypothesis I have eliminated a large number of animals from the area, but not the 10 adult animals per breathing hole. I have no idea what constitutes a crowded breathing hole under normal conditions, but at the experimental sites where I have released seals (as described in Chapter 5), 2 is a crowd. Based on a bias from my own experiments, I suspect that 10 is certainly a crowd under circumstances of tight ice conditions and dispersed breathing holes. So I suggest that some of the more aggressive animals exercise crowd control and induce others to leave as winter sets in and sea ice forms and thickens in the Sound. This reduces the competition for the available resources that the seals can reach from a particular breathing hole, and it reduces congestion at the breathing hole (most breathing holes are so small that only one seal can get into them at a time). In conclusion, I tend to agree with Smith's conservatism, although not with his numbers. There are fewer seals and they are widely dispersed in the Sound during the winter.

So how many Weddell seals are there? As you can see, most information about Weddell seals comes from populations in McMurdo Sound, which might be quite different from those throughout the rest of Antarctica. The need to know the population sizes of these and other animals is important. So scientists persevere in their efforts to find out. As a general rule, the methods are primitive, even though some of the tools may be quite elaborate.

A common procedure for a local population like the seals in Mc-Murdo Sound is to mark the animals, or a percentage of them, and later count their occurrence in the population. It is sort of the black-sheep-in-the-flock rationale; except the marking methods have some problems. Branding, once a fairly common procedure, may increase the mortality of the animals, especially of pups. I have seen brands on adults that we know were done twelve to fifteen years earlier and they have never healed properly – there is still an open, seeping wound in the center of the brand scar! Another much-used method is placing a tag on the front or hind flippers. Unfortunately, especially with pups, the most frequently tagged animals, there is a loss that is unknown but may be substantial. Consequently, the difficulties of estimating numbers of animals, their recruitment, and their mortality are compounded. It is somewhat like using as your marked animals some black sheep, of which an unknown number turn white as they mature, or have a much higher rate of congenital defects than white sheep and die sooner.

On the much grander scale of estimating populations over thousands of square kilometers, or even the world population, which for Weddell seals means the antarctic and subantarctic populations, the procedures become even shakier. Usually figures for estimates are obtained from shipboard or aerial counts, or a combination of both. A measured sector of the ocean is traversed, and all the seals seen are counted. Then it is assumed that the animals are distributed in a similar fashion over the entire area, and the total population of large regions is calculated. Another problem is that much of the time they are under water and out of sight, so some factor has to be assumed to account for these individuals. You might consider that it is like trying to estimate the population of New York City by an aerial survey where you count the number of people on the beaches, in the parks, or anywhere that they are visible within the city. The timing of the count would be crucial. An estimate based on data from July 4 would be quite different from a survey made on December 21.

The census accuracy can be improved considerably with knowledge of the habits of the animals under study. Much of the work of population ecologists is devoted to this area, and it is continually helping them to improve their data base.

So how many Weddell seals are there? At present, only God knows, but we must try to find out in order to gain some assessment of what effect we are having on local populations when we establish bases and introduce the host of perturbations that necessarily accompany such activities; or on world populations with the broader effects of resource exploitation.

Based on numerous censuses of various sorts made continuously over nearly twenty years, the McMurdo Sound population from Cape Royds to Cape Armitage is usually between 2,000 and 3,000 animals in late summer. However, this must be qualified, because the Mc-Murdo Sound population seems to be declining since the mid-1960s, when the highest counts of 2,960 animals were made. In January 1979, only 811 seals were counted. Even pup counts, which may be a more reliable index, show this downward trend from a peak number of 767 in 1967 to only 250 in 1978.

The 500 to 800 miles of near-shore coastline of the Western Ross Sea from Cape Adare South to McMurdo Sound may have a population of 50,000 animals. This estimate is based upon a combination aerial and shipboard census during 1967 and 1968 (Stirling, 1968). The approximate number of seals counted was 4,528, which is quite a good fraction of the total estimate compared to many marine mammal censuses.

The estimate of Weddell seals in the Weddell Sea, based on counts made from January to March 1968, was 593,700 (Siniff, Cline and Erickson, 1975). The number of seals counted was 82. The estimate for February 1969 in the same area by the same investigators was 92,900. No world estimates were attempted, but they suggested that it must be much higher than the often-quoted 200,000 to 500,000 given by Vic Scheffer in his well-known book *Seals, Sea Lions and Walruses*, (Scheffer, 1958). Most recently (1977), the highly respected population ecologist Richard Laws has "hesitatingly" suggested that the antarctic "stock" size of Weddell seals is perhaps 730,000 (Laws, 1977). In short, based on present techniques, it is probably more than a half million and fewer than 1 million.

Like most marine mammals, particularly adults, there is little known about the mortality of Weddell seals. Most of the information available is from data obtained at McMurdo Sound. These data must be interpreted cautiously because the Sound may have unique conditions subjecting seals to quite different hazards from those affecting the general population around the rest of the continent.

Typical of McMurdo Sound, but perhaps less typical elsewhere, is the necessity for the seals to open and maintain holes by reaming with their teeth. In fast ice areas like McMurdo, this reaming is essential and the brunt of such activities is borne by the upper canines and second incisors. The worn teeth of older animals show the price paid for countless assaults on ice holes. Eventually, the wear will cause a break into the pulp cavity. Once this occurs the tooth, upper jaw, palate and nasal cavity are subject to the ravages of eroding infections (Stirling, 1969).

The breathing holes are a catalyst to intraspecific aggression. Much of the fighting among seals seems to be a direct result of efforts by seals to control access to breathing holes. The cuts and tears sustained in these fights may not be important, but the constant harassment of old or young animals may hasten their demise. Prevented from feeding or resting properly, the physical conditions of these animals deteriorate.

An important biological factor causing mortality is predation. The seals that remain in the Sound are rather well separated physically from the only two natural predators in the Antarctic: the killer whale and the leopard seal. These two species appear in the area only in the late summer when the fast ice breaks up. This breaking up may occur at the time the pups are dispersing; the youngsters tend to remain near open water or loose ice areas, and killer whales very likely take their toll. Along the continental margins to the north, both pups and adults may be more subject to such predation. Indeed, spectacular catches, where the killer whales actually tilt an ice floe and cause the seal to slip from this sanctuary into their grasps, have been observed. The leopard seal is probably of little importance as a predator, and then only of young animals.

In the Sound, the greatest predator of adult seals is man. This is a recent phenomenon, and it is still not understood how much effect this unnatural condition has on the population. It began at the turn of the century when the first expedition overwintered at McMurdo Sound. The members of the expedition took seals to feed both men and dogs. This ended in about 1914. Then, in 1956, both the United States and New Zealand established bases on Ross Island. Both countries used dog teams that year and New Zealand killed 350 seals (Stirling, 1971a). The United States did not record the number it took. By 1958 or 1959 the United States no longer used dogs, and ceased killing seals for food. New Zealand has taken progressively fewer seals; 105 in 1957 and 1958, down to 60 by 1967 and 1968. At the present time (1977), New Zealand still kills about 40 adult seals a year.

The effect of parasites is not known, but they probably take some toll. Seals are infested with a body louse, *Antarctophthirus ogmorhini*. This louse is most abundant on the pup and yearling animals and is probably more of a nuisance than a cause of death. On the other hand, almost all seals are infested heavily with helminths – parasitic gastrointestinal worms. Some of the most abundant are cestodes (of the genera *Diphyllobothrium,* and *Glandicephalus*), a trematode (*Ogogaster*), and a nematode (*Contracaecum*). Some of these are at times so abundant that blockage of the bile duct and perhaps other passageways seems possible. The end result of such obstructions could be death. Several adult seals are found dead on the ice each year with no obvious causes. Some of these may be due perhaps to a critical imbalance between host and parasite.

Some physical features of the environment can be hazardous. One that comes to mind immediately is the few breathing holes over vast areas of ice. However, as I mentioned in Chapter 8, I believe it is very rare that seals get so lost while swimming under the ice that they drown because of failure to find an exit. Ice of another sort is a distinct hazard. Often the preferred haul-out sites are along tidal cracks adjacent to high cliffs that have large ice and snow cornices. From time to time, these cornices collapse and any seals caught below are crushed. One such icefall occurred in 1964 at Arrival Heights, where 15 to 30 seals regularly slept in the afternoon. The massiveness of the rubble heap prevented me from ascertaining whether any seals were killed. If the icefall occurred in late afternoon, the preferred resting period, I suspect many were.

The Erebus icefall pupping area seems to be an especially dangerous place for pups. In one survey, I found a dead pup under a collapsed ice cornice and another in one of many narrow ice cracks. It could not climb out and starved to death.

In a survey conducted from 1969 to 1971 in the relatively safe area of Hutton Cliffs, 162 pups were born and 10 died in the colony (Kaufman, Siniff and Reichle, 1975): six of unknown causes next to their mothers, one by an ice cornice fall, two by being abandoned, and one because it was unable to get out of the water and probably died of hypothermia.

During the summers of 1966 to 1967 and 1967 to 1968, Ian Stirling found seven dead adult seals (Stirling, 1971b). Their ages ranged from 7 to 16 years. These were four females and three males. Based on tagging and resighting analysis, he estimated that the minimum annual survival rate for adult males was 76 percent, and for adult females 86 percent. He calculated also that at least 46 percent of the female pups born survive to their first breeding. Pups, for which his mortality estimates are most accurate, have a survival rate from birth to weaning of 95 percent.

Suggested reading

Gilbert, J. R., and A. W. Erickson. 1977. Distribution and abundance of the Pacific sector of the southern ocean. In *Adaptations Within Antarctic Ecosystems, Proceedings of the 3rd Special Committee on Antarctic Research Symposium on Antarctic Biology*, pp. 703–40, ed. G. A. Llano. Washington, D.C.: Smithsonian Institution.

10

Future prospects

What is past is prologue.
Shakespeare, The Tempest

During the heroic era of the late 1800s and early 1900s, the killing of seals was a necessity if polar exploration were to go forth. Expeditions lasted two or three years, and it was impossible to bring enough fresh meat for the entire time. Also, dog teams were frequently brought for overland transportation and they required a considerable amount of meat. Seals and penguins were often a viable emergency ration. In several instances, exploratory parties ran out of supplies and had to rely on local resources. Two of the most dramatic examples are Scott's northern party expedition in 1912 and Shackleton's entire expeditionary force in 1915.

On January 8, 1912, six men disembarked the ship *Terra Nova* in Terra Nova Bay on the northern coast of the Ross Sea. The ship was unable to return for them, and they had to overwinter unexpectedly in a cave on Inexpressible Island. They did not have nearly enough food supplies, and they survived because of the penguins and seals they killed. On November 7, 1912, they reached the main camp after about one month's travel over the coastal sea ice.

Perhaps the most dramatic polar experience in history was that of Sir Ernest Shackleton and his men. The ship *Endurance* was beset in the Weddell Sea and on October 24, 1915, the hull-crushing forces of the sea ice caused the ship to sink. For nearly one year, until August 28, 1916, the party of fifty six men lived on the drifting ice of the Weddell Sea and then later on bleak Elephant Island. If they had not been able to catch seals and penguins, survival would have been impossible.

At present, none of the seals killed in Antarctica are used as food for man. Most of them are consumed by dogs and the south polar skuas that scavenge the cache. Some of the seal meat is used for bait in fish and invertebrate traps, and some animals are sacrificed for ecological and anatomical studies.

Between 1964 and 1969, the number of Weddell seals killed by all scientific programs in the Antarctic was 893. Such a small number of seals relative to their total population is thought to have little or no effect. In most cases this is true, but, because of the fidelity of Weddell seals to certain coastal areas, local populations are especially vulnerable to killing and other man-induced disturbances – such as periodic harassment during outings by station personnel and noisy modes of transportation by land and air. There are no well-documented studies of such effects, but there is an ominous historical comparison. In E. A. Wilson's account (Wilson, 1907) of the seals in the McMurdo Sound area, he noted that the seal-pupping area near Pram Point was the largest of any other known as far north as Cape Royds. Today no seals pup at Pram Point. Whether the abandonment of this area as a pupping site has been a result of the development of two major stations, New Zealand's Scott Base and the United States' McMurdo Station, is unknown. If the disappearance of the Pram Point pupping site is due to the presence of the bases, this is unfortunate. So is the continued collecting in the area because, as I have attempted to point out in this book, the seals of McMurdo Sound are unique and the apparent decline in their numbers may jeopardize future studies of them.

Pollution may be another serious matter on the local level. Litter and sewage are perhaps the most important logistic problems of a permanent base. Our technology for bringing supplies into an area is superb. But our technology for waste disposal is still in the Stone Age, and archaeologists of future years will probably explore our middens as present-day archaeologists do those of early aborigines. At coastal stations, garbage and raw sewage are dumped directly into the sea. Ecologists studying the shallow water-bottom fauna near McMurdo Station have noted that some areas are carpeted with litter. This litter includes everything imaginable, such as fuel lines, barrels, rope, tractors, parts of airplanes, and, of course, most abundant of all, beer cans. These biologists also noted that decomposition appears to be very slow in these frigid waters.

In the course of ship movements in antarctic waters and the refueling of bases, oil spills have occurred. These spills represent local contaminations and do not present a serious general problem to the wildlife populations of Antarctica. At present, the activities of man in Antarctica represent only local pockets of degradation where he is fouling mainly his own nest. When his interest in utilization of antarctic resources becomes much greater, a different story will unfold. Unfortunately, the first chapter in that story may be near at hand.

In the past few years, there has been speculation about whether the harvesting of antarctic seals could be profitable. Thus far, the major protection of these seals has been their remoteness and the fact that the principal species, the crabeater seal, is not concentrated in specific dependable localities, but is scattered throughout the pack ice. Indeed, the only antarctic seal that concentrates in numbers is the Weddell seal, which, relative to the total number of crabeater seals, is not numerous. Present-day technology may have eliminated the major deterrents to antarctic sealing. If so, the only protection for the seals is the Antarctic Treaty.

There are twelve signatory nations to this treaty: Argentina, Australia, Belgium, Britain, Chile, France, Japan, New Zealand, Norway, South Africa, the United States, and the Soviet Union. The treaty accords dedicate Antarctica and all lands south of the Sixtieth Parallel to peaceful purposes. The treaty does not cover the high seas; thus the Antarctic Ocean, the mother of all biological resources in the region, is unprotected by international agreement.

Does our past record of exploitation of the high seas leave us with much optimism for the future prospects of antarctic seals if commercial hunting begins? It does not.

In the 1770s, Cook was searching for the supposed southern continent, *Terra Incognita*. He never found it, but in 1775 he discovered South Georgia Island. Included in his report on this subantarctic island was news of the enormous population of seals and fur seals, *Arctocephalus gazella*. This had the effect of essentially opening a Pandora's box in Antarctica that has yet to be closed. Sealers in numbers came for the pelts of the fur seals and the oil of sea lions, *Otaria flavescens*, elephant seals, *Mirounga leonina*, and penguins. By 1800, all the subantarctic islands were virtually depopulated of large mammals, and the sealers were searching further south. In 1819, sealers began their slaughter of fur seals in the South Shetland Islands. By 1823, over 320,000 fur seals had been killed and the industry collapsed for lack of animals (Stonehouse, 1972). In the little time since

Cook's discovery, men working with primitive tools and ships had exterminated millions of fur seals that inhabited the subantarctic and antarctic islands – islands that they had no idea existed until they began to hunt.

Slightly more than half a century later, the slaughter began afresh; this time it was the whales. By the early 1900s, antarctic whaling was in full swing. From 1904 to 1930, the catch steadily increased; a record 40,000 whales were taken in 1930. There was a brief respite during the Depression, but by the late 1930s the killing was above previous levels; 46,000 whales were taken in the 1937 to 1938 season. In the late 1930s and early 1940s, the killing of men took precedence over the whales, although according to some many whales fell victim to military maneuvers and training. After the war, it was obvious that the killing required some regulation, and the International Whaling Commission was established. Seventeen nations participated in this organization, including all those with major whaling fleets. The commission became very effective in predicting the stocks of whales and the expected catches, based on previous effort. Unfortunately, it had no enforceable regulatory powers and the slaughter continued unchecked. The population declined rapidly and the industry began to contract. In 1962, South Africa stopped pelagic whaling. In 1963, Britain sold its fleet to Japan. The Netherlands stopped whaling in 1964, followed by Norway in 1968. Only Japan and the Soviet Union are still in the pelagic whaling business today. Small numbers of whales are also taken by some countries that carry on shore-based whaling operations.

In 1965, the blue whale, *Balaenoptera musculus*, was given "full" protection. In the previous sixty years, 325,000 blue whales were taken from antarctic waters. The stock is now so low that there is much uncertainty about whether the population decline will stop. In the 1950s, when the number of blue whales was decreasing sharply, the hunting effort was directed more toward the fin whale, *Balaenoptera physalus*; during that period, 240,000 were killed. The fin whale population has now dropped to a point where more and more of the hunting effort must be focused upon the smaller whales. At present, five of nine species of whale previously hunted are considered to be endangered species. One or two others may soon join that list.

The antarctic seal resources are unique. Except for a few years of sealing in the early 1890s, in which a total of 132,500 seals were taken, and for a small expedition in 1964, in which 861 seals were killed (Oritsland, 1970), there has been no exploitation of these animals. At the present time, efforts are being made to estimate the stocks to determine the numbers that could be taken without harming the population. In 1972, a conference was held to suggest catch limits before an industry was begun. This was a unique event in the history of wildlife harvests. Twelve nations, the signatories of the Antarctic Treaty, participated in the conference. The animal catches suggested were 175,000 crabeater seal, 12,000 leopard seal, and 5,000 Weddell seal. Since that time, the figures have been enlarged to 200,000 crabeater seal, 15,000 leopard seal, and 10,000 Weddell seal.

If sealing begins in earnest in Antarctica, what are the future prospects for the Weddell seal? I am afraid that if the first sealing expeditions are a financial success, a new chapter in an old story will begin. Regulations will be ignored or become rationalized victims of the profit motive. The result will be overexploitation and a rapid decline in the seals. The principal seal that will support the industry will be the crabeater, but in the process of their slaughter the other species will not be ignored. The Weddell seal will be especially vulnerable because of its tendency to congregate in the summer on stable fast ice and because it takes no notice as other members of a group are shot. As a result, discrete, local populations will be eliminated.

And for what purpose will this great resource of seals be used? Not likely to feed the starving masses we so often hear and read about, but rather to put furs on the backs of fashion-conscious men and women and to feed the exploding population of cats and dogs in North America and Western Europe!

What will we have sacrificed in order to accomplish this? We will see the demise of the last pristine environment and untouched animal population in the world. Not even aboriginal man hunted these resources. The idea that there is a great stronghold in the world where animals other than man are the masters has both practical and aesthetic attributes.

The seals and birds of Antarctica are the only great population of large animals that do not live in national parks or special reserves, and are unmanaged. The limits of their populations and distribution are set by natural laws, not by man. In essence, they are our last chance of retaining a fauna that can be studied in a wholly natural state and self-contained, relatively simple ecosystem. There is still much of great value to be learned from such animal populations.

Furthermore, these animals represent a source of protein that is essentially in reserve. They live in a relatively uncontaminated environment, so that unforeseen population declines due to pollution are unlikely in the near future. Consequently, this population should remain intact at least until we finally arrive at a sensible international solution to the use of the sources of the sea.

Aesthetically, the possibilities are as numerous as the thoughts of human beings. Large, wild animal populations have many effects on human beings. The great attraction of the animal parks of East Africa is the most notable. What is there about seeing vast herds of animals in their natural surroundings that appeals to people? The experience is difficult to define, but there is no doubt that it deeply touches most.

If the antarctic seals go, the material effect on the people of the world will be insignificant. For a few, it will cause a profound emotional disturbance, but certainly not as many people will be disturbed as those who are concerned about the slaughter of the great whales. But for everyone, whether they realize it or not, there will be a great loss. Consider it this way. These seal populations represent one of the great natural wonders of the world, comparable aesthetically to the works of Bach, Beethoven, and Mozart. These composers need never have existed. If they had not, it would make little difference to us materially, as it made little difference to people before 1685. But without their works our world would be more stark, and certainly we would not now think of expunging their compositions from the musical record. Similarly, our apparent survival or creature comforts are not affected by the existence or absence of many of the world's wonders. But without them we might be less awe-struck, less concerned, and perhaps more willing to plunge even faster toward the environmentally uniform world that human beings in general seem intent upon creating as we "progress" in our "development" of the environment.

References

Chapter 1. McMurdo Sound

Dayton, P. K., G. A. Robilliard, and R. T. Paine. 1970. Benthic faunal zonation as a result of anchor ice at McMurdo Sound, Antarctica. In *Antarctic Ecology*, vol. 1, pp. 244–58, ed. M. W. Holdgate, London: Academic Press.

Raymond, J. A. 1975. Fishing for Antarctica's largest fish, the antarctic cod. *Marine Technology Society Journal* 9:32–5.

Ruud, J. T. 1965. The ice fish. *Scientific American* 213:108–14.

Chapter 2. The Weddell seal

Bertram, G. C. L. 1940. The biology of the Weddell and crabeater seals. British Graham Land Expedition 1934–7, *Scientific Reports* 1:1–139. London: British Museum of Natural History.

Lindsey, A. A. 1937. The Weddell seal in the Bay of Whales, Antarctica. *Journal of Mammalogy* 18:127–44.

Sapin-Jaloustre, J. 1952. Les phoques de Terre Adélie. *Mammalia* 16:179–212.

Weddell, J. A. 1971. *A Voyage Towards the South Pole–Performed in the Years 1822–1824.* Newton Abbot, Great Britain: David and Charles Reprints.

Chapter 3. Breeding, birth, and growth

Elsner, R., D. D. Hammond, D. M. Denison, and R. Wyburn. 1977. Temperature regulation in the newborn Weddell seal *Leptonychotes weddelli*. In *Adaptations within Antarctic Ecosystems*, pp. 531–40, ed. G. A. Llano. Washington, D. C.: Smithsonian Institution.

Kaufman, G. W., D. B. Siniff, and R. Reichle. 1975. Colony behavior of Weddell seals, *Leptonychotes weddelli*, at Hutton Cliffs, Antarctica. *Rapports Procès Verbaux Réunions, Conseil International Exploration Mer.* 169:228–46.

Kooyman, G. L. 1969. The Weddell seal. *Scientific American* 221:100–6.

Kooyman, G. L., and C. M. Drabek. 1968. Observations on milk, blood and urine constituents of the Weddell seal. *Physiological Zoology* 41:187–93.

Lindsey, A. A. 1937. The Weddell seal in the Bay of Whales, Antarctica. *Journal of Mammalogy* 18:127–44.

References

Chapter 4. Cold

Davydov, A. F., and A. R. Makarova. 1964. Changes in heat regulation and circulation in newborn seals on transition to aquatic form of life. Translation Supplement No. 4, Part II, *Federation Proceedings* 24:T563–6.
Elsner, R., D. D. Hammond, D. M. Denison, and R. Wyburn. 1977. Temperature regulation in the newborn Weddell seal *Leptonychotes weddelli*. In *Adaptations within Antarctic Ecosystems*, pp. 531–40, ed. G. A. Llano. Washington, D. C.: Smithsonian Institution.
Grav. H. J., A. S. Blix, and A. Pasche. 1974. How do seal pups survive birth in arctic winter? *Acta Physiologica Scandinavica* 92:427–9.
Hart, J. S., and L. Irving. 1959. Energetics of harbor seals in air and in water with special considerations of seasonal changes. *Canadian Journal of Zoology* 37:447–57.
Irving, L., and J. S. Hart. 1957. The metabolism and insulation of seals as bareskinned mammals in cold water. *Canadian Journal of Zoology* 35:497–511.
Kerslake, D. McK. 1972. *The Stress of Hot Environments*. Cambridge: Cambridge University Press.
Kooyman, G. L., and D. H. Kerem, W. B. Campbell, and J. J. Wright. 1973. Pulmonary gas exchange in freely diving Weddell seals, *Leptonychotes weddelli*. *Respiration Physiology* 17:283–90.
Miller, K., M. Rosenmann, and P. Morrison. 1976. Oxygen uptake and temperature regulation of young harbor seals (*Phoca vitulina richardi*) in water. *Comparative Biochemistry and Physiology* 54A:105–7.
Scholander, P. F., V. Walters, R. Hock, and L. Irving. 1950. Body insulation of some arctic and tropical mammals and birds. *Biological Bulletin* 99:225–36.

Chapter 5. Diving behavior

Heezen, B. C. 1957. Whales entangled in deep-sea cables. *Deep-Sea Research* 4:105–15.
Kooyman, G. L. 1965. Techniques used in measuring diving capacities of Weddell seals. *Polar Record* 12:391–4.
Kooyman, G. L., E. A. Wahrenbrock, M. A. Castellini, R. W. Davis, and E. E. Sinnett. 1980. Aerobic and anaerobic metabolism during voluntary diving in Weddell seals: evidence of preferred pathways from blood chemistry and behavior. *Journal of Comparative Physiology B, Biochemical, Systemic and Environmental Physiology* (in press).
Lockyer, C. 1977. Observations on diving behaviour of the sperm whale *Physeter catodon*. In *A Voyage of Discovery*, pp. 591–609, ed. M. Angel. New York: Pergamon Press.
Scholander, P. F. 1940. Experimental investigations on the respiratory function in diving mammals and birds. *Hvalradets Skrifter Norske Videnskaps-Akad* (Oslo) 22:1–131.

Chapter 6. Physiology of diving

Andersen, H. T. 1966. Physiological adaptations in diving vertebrates. *Physiological Reviews* 46:212–43.
Bert, P. 1870. De l'acte du plonger chez les mammiferes et chez les oiseaux, Trentieme leçon. In *Leçons sur la physiologie comparée de la respiration*. Paris: J.-B. Bailliere et Fils.
Elsner, R., D. L. Franklin, R. L. Van Citters, and D. W. Kenney. 1966. Cardiovascular defense against asphyxia. *Science* 153:941–9.

References   129

Graham, S. F. 1967. Seal ears. *Science* 155:489.
Halasz, N. A., R. Elsner, R. S. Garvie, and G. T. Grotke. 1974. Renal recovery from
ischemia: a comparative study of harbor seal and dog kidneys. *American Journal
of Physiology* 227:1331-5.
Hunter, W. L., and P. B. Bennett. 1974. The causes, mechanisms and prevention of
the high-pressure nervous syndrome. *Undersea Biomedical Research* 1:1-28.
Irving, L. 1964. Comparative anatomy and physiology of gas transport mechanisms.
In *Handbook of Physiology, Section III: Respiration*, vol. 1, pp. 177-212, ed. W. O.
Fenn and H. Rahn. Washington, D.C.: American Physiological Society.
Kooyman, G. L. 1972. Deep diving behaviour and effects of pressure in reptiles,
birds and mammals. *Symposia of the Society for Experimental Biology: The Effects
of Pressure on Organisms*, vol. 26, pp. 295-311, ed. M. A. Sleigh and A. G. Mac-
Donald. Cambridge: Cambridge University Press.
- 1973. Respiratory adaptations in marine mammals. *American Zoologist* 13:457-68.
- 1975. A comparison between day and night diving in the Weddell seal. *Journal of
Mammalogy* 56:563-74.
Kooyman, G. L., and W. B. Campbell. 1972. Heart rates in freely diving Weddell
seals, *Leptonychotes weddelli*. *Comparative Biochemistry and Physiology* 43A:31-7.
Kooyman, G. L., D. D. Hammond, and J. P. Schroeder. 1970. Bronchograms and
tracheograms of seals under pressure. *Science* 169:82-4.
Kooyman, G. L., J. P. Schroeder, D. M. Denison, D. D. Hammond, J. J. Wright,
and W. P. Bergman. 1972. Blood N$_2$ tensions of seals during simulated deep
dives. *American Journal of Physiology* 223:1016-20.
Kooyman, G. L., E. A. Wahrenbrock, M. A. Castellini, R. W. Davis, and E. E. Sinnett. 1980. Aerobic and anaerobic metabolism during voluntary diving in Wed-
dell seals: evidence of preferred pathways from blood chemistry and behavior.
*Journal of Comparative Physiology B, Biochemical, Systemic and Environmental Phys-
iology* (in press).
Odend'hal, S., and T. C. Poulter. 1966. Pressure regulation in the middle ear cavity
of sea lions: a possible mechanism. *Science* 153:768-9.
Paulev, P. 1965. Decompression sickness following repeated breath-hold dives.
Journal of Applied Physiology 20:1028-31.
Schaefer, K. E., R. D. Allison, J. H. Dougherty, C. R. Carey, R. Walker, F. Yost,
and D. Parker. 1968. Pulmonary and circulatory adjustments determining the
limits of depths in breathhold diving. *Science* 162:1020-3.
Scholander, P. F. 1940. Experimental investigations on the respiratory function in
diving mammals and birds. *Hvalradets Skrifter Norske Videnskaps-Akad* (Oslo)
22:1-131.
- 1963. The master switch of life. *Scientific American* 209:92-106.
- 1964. Animals in aquatic environments: diving mammals and birds. In *Handbook
of Physiology, Section IV: Adaptation to the Environment*, pp. 729-39. Washington,
D. C.: American Physiological Society.

Chapter 7. Food habits and energetics

Bertram, G. C. L. 1940. The biology of the Weddell and crabeater seals. *British Graham Land Expedition 1934-1937*, Scientific Report 1:1-139, London: British Museum of Natural History.
Calhaem, I., and D. A. Christoffel. 1969. Some observations of the feeding habits of
a Weddell seal and measurements of its prey *Dissostichus mawsoni*, at McMurdo
Sound, Antarctica. *New Zealand Journal Marine and Freshwater Research* 3:181-90.
Davydov, A. F., and A. R. Makarova. 1964. Changes in heat regulation and circula-

tion in newborn seals on transition to aquatic form of life. *Translation Supplement No. 4, Part II, Federation Proceedings* 24:T563–6.

Dearborn, J. H. 1965. Food of Weddell seals at McMurdo Sound, Antarctica. *Journal of Mammalogy* 46:37–43.

Elsner, R., D. D. Hammond, D. M. Denison, and R. Wyburn. 1977. Temperature regulation in the newborn Weddell seal *Leptonychotes weddelli*. In *Adaptations within Antarctic Ecosystems*, pp. 531–40; ed. G. A. Llano. Washington, D. C.: Smithsonian Institution.

Goldspink, G. 1977. Energy cost of locomotion. In *Mechanics and Energetics of Animal Locomotion*, pp. 153–67, ed. R. Mcn. Alexander and G. Goldspink. New York: Wiley.

Hart, J. S., and L. Irving. 1959. Energetics of harbor seals in air and in water with special consideration of seasonal changes. *Canadian Journal of Zoology* 37:447–57.

Kooyman, G. L., D. H. Kerem, W. B. Campbell, and J. J. Wright. 1971. Pulmonary function in freely diving Weddell seals, *Leptonychotes weddelli*. *Respiration Physiology* 12:271–82.

– 1973. Pulmonary gas exchange in freely diving Weddell seals, *Leptonychotes weddelli*. *Respiration Physiology* 17:283–90.

Kruse, D. H. 1975. Swimming metabolism of California sea lions, *Zalophus californianus*. M.Sc. thesis, San Diego State University.

Miller, K., M. Rosenmann, and P. Morrison. 1976. Oxygen uptake and temperature regulation of young harbor seals (*Phoca vitulina richardi*) in water. *Comparative Biochemistry and Physiology* 54A:105–7.

Schmidt-Nielsen, K. 1972. Locomotion: Energy cost of swimming, flying and running. *Science* 177:222–8.

Chapter 8. Under-ice orientation

Elsner, R., G. L. Kooyman, and C. M. Drabek. 1970. Diving duration in pregnant Weddell seals. In *Antarctic Ecology*, pp. 477–82, ed. M. W. Holdgate. New York: Academic Press.

Kooyman, G. L. 1975. A comparison between day and night diving in the Weddell seal. *Journal of Mammalogy* 56:563–74.

Kooyman, G. L., and M. A. Castellini. Unpublished observation.

Schevill, W. E., and W. A. Watkins. 1971. Directionality of the sound beam in *Leptonychotes weddelli* (Mammalia: Pinnipedia). In *Antarctic Pinnipedia, Antarctic Research Series* 18, pp. 163–8, ed. W. H. Burt. Washington, D. C.: American Geophysical Union.

Smith, M. S. R. 1965. Seasonal movements of the Weddell seal in McMurdo Sound, Antarctica. *Journal of Wildlife Management* 29:464–70.

Chapter 9. Distribution, abundance, and mortality

Erickson, A. W., and R. J. Hofman. 1974. Antarctic seals. In *Antarctic Mammals*, Antarctic Map Folio Series, Folio 18. New York: American Geographical Society.

Kaufman, G. W., D. B. Siniff, and R. Reichle. 1975. Colony behavior of Weddell seals, *Leptonychotes weddelli*, at Hutton Cliffs, Antarctica. Rapp. *Procès Verbaux Réunions, Conseil International Exploration Mer.* 169:228–46.

Kooyman, G. L. 1968. An analysis of some behavioral and physiological characteristics related to diving in the Weddell seal. In *Biology of the Antarctic Seas*, Antarctic Research Series, vol. 3, pp. 227–61, ed. W. L. Schmitt and G. A. Llano. Washington, D. C.: American Geophysical Union.

Laws, R. M. 1977. The significance of vertebrates in the antarctic marine ecosystem. In *Adaptations within Antarctic Ecosystems, Proceedings of the 3rd Special Committee on Antarctic Research Symposium on Antarctic Biology,* pp. 411–38, ed. G. A. Llano. Washington, D. C.: Smithsonian Institution.

Scheffer, V. B. 1958. *Seals, Sea Lions and Walruses.* Stanford, Calif.: Stanford University Press.

Schevill, W. E., and W. A. Watkins. 1971. Directionality of the sound beam in *Leptonychotes weddelli* (Mammalia: Pinnipedia). *Antarctic Pinnipedia,* Antarctic Research Series 18, pp. 163–8, ed. W. H. Burt. Washington, D. C.: American Geophysical Union.

Siniff, D. B.,. D. R. Cline, and A. W. Erickson. 1970. Population densities of seals in the Weddell Sea, in 1968. In *Antarctic Ecology,* vol. 1, pp. 377–94, ed. M. W. Holdgate. London: Academic Press.

Siniff, D. B., D. P. DeMaster, R. J. Hofman, and L. L. Eberhardt. 1977. An analysis of the dynamics of a Weddell seal population. *Ecological monographs* 47:319–35.

Smith, M. S. R. 1966. Studies on the Weddell seal in McMurdo Sound, Antarctica. Ph.D. thesis, University of Canterbury, New Zealand.

Stirling, I. 1969. Distribution and abundance of the Weddell seal in the western Ross Sea, Antarctica. *New Zealand Journal of Marine and Freshwater Resources* 3:191–200.

– 1969. Tooth wear as a mortality factor in the Weddell seal, *Leptonychotes weddelli. Journal of Mammalogy* 50:559–65.

– 1971a. Population aspects of Weddell seal harvesting at McMurdo Sound, Antarctica. *Polar Record* 15:653–67.

– 1971b. Population dynamics of the Weddell seal *(Leptonychotes weddelli)* in McMurdo Sound, Antarctica 1966–1968. *Antarctic Research Series* 18:141–61.

Thomas, J., V. Kuechle, D. DeMaster, E. Birney, and J. Eldridge. 1978. Colonial behavior of the Weddell seal in eastern McMurdo Sound. *Antarctic Journal* 13:159–60.

Wilson, E. A. 1907. Weddell's seal, Mammalia. In *British National Antarctic Expedition Report 1901–1904, Natural History* 2:1–66, London: British Museum.

Chapter 10. Future prospects

Laws, R. M. 1972. Seals and birds killed and captured in the antarctic treaty area, 1964–1969. *Polar Record* 16:343–5.

Oritsland, T. 1970. Sealing and seal research in the southwest Atlantic pack ice, September-October, 1964. In *Antarctic Ecology,* vol. 1, pp. 367–76, ed. M. W. Holdgate. London: Academic Press.

Stonehouse, B. 1972. *Animals of the Antarctic – The Ecology of the Far South.* New York: Holt, Rhinehart and Winston.

Wilson, E. A. 1907. Weddell's seal, Mammalia. In *British National Antarctic Expedition 1901–1904,* Natural History, 2:1–66. London: British Museum.

Index

Page numbers followed by f refer to figures; those followed by t refer to tables